I0813767

THE

HYDRANGEA

A REAPPRAISAL

Maurice Foster

THE HYDRANGEA

A REAPPRAISAL

THE CROWOOD PRESS

CONTENTS

PART I HISTORY AND ORIGINS

CHAPTER 1

A 'POTTED' HISTORY

In 1736, John Bartram, son of a Quaker farmer migrant and known by some as the father of American botany, collected plants and seeds of *Hydrangea arborescens* from the Appalachian mountains down the eastern side of what was soon to become the United States. These finds were among many baskets of plants that he sent to a London merchant, Peter Collinson, keen to acquire new plants from the New World.

The first introduction, *H. arborescens*, sent to England from North Carolina.

From the small white-flowered corymbs of these plants it is clear why, although the first hydrangea introduced to Europe was an American, it was not until the introduction of the first living plants from Japan and China 50 years later that interest in hydrangeas as a genus began to stir.

The impact of *Hydrangea macrophylla* from Japan

In 1789, a plant called *Hydrangea hortensis* was brought to Kew through the agency of Sir Joseph Banks from a cultivated plant in China, although originally from Japan.

The eighteenth century was a time of a significant exchange of plants between China and Japan, and this form of hydrangea gained attention for its showy globose flower head, typical of the now ubiquitous and varied cultivars of the Japanese maritime species *H. macrophylla*. Nearly two centuries later, the original import was christened 'Joseph Banks' by Michael Haworth-Booth (an English plantsman and nurseryman, and author of the first significant British book on hydrangeas, in 1950). This original cultivar is still grown today where conditions are mild: old plants can still be seen, for example, flourishing in cottage gardens in the soft maritime climate of the south

OPPOSITE PAGE: *H. macrophylla* 'Le Cygne' (Henri Cayeux, 1919).

Cornish coast of England. Indeed, Howarth-Booth discovered its origin when he found on the Isle of Wight that it had reverted freely to a wild form of *H. macrophylla*, whence 'Joseph Banks' had clearly arisen as a branch sport. He renamed this in 1950 as a new species, *H. maritima*, with the cultivar name of 'Sea Foam'. It was later deemed illegitimate as a new species.

Two species of Japanese hydrangeas had previously been discovered by Carl Thunberg, a Swedish physician, botanist and representative of the Dutch East India Company in Japan, although he identified and described these two new plants as viburnums. Japan was closed to foreigners in the 1770s, but the Dutch East India Company was allowed to trade as a special concession by the Japanese, as unlike other organisations it had not sent unwelcome proselytising missionaries into Japanese society.

The authorities allowed Thunberg to live on an artificial island called Deshima in Nagasaki bay, and he was permitted one visit to the mainland each year to pay homage to the Emperor in Tokyo, but was not allowed to collect plants on these trips, so most of his information about the Japanese flora came from local foraging. He evidently kept goats, and his servants collected fodder on the mainland. Thunberg was able to examine this material, and is said to have discovered his two hydrangeas among it, described in his *Flora Japonica* in 1784.

If this account of the novel source for Thunberg's new plants is overly imaginative, another possibility is that they were simply brought to him by Japanese friends who appreciated his passion for plants. In any event, he named the two

The first Japanese mophead (*H. macrophylla*) from Sir Joseph Banks.

H. macrophylla 'Sea Foam', considered the original source of 'Joseph Banks'.

Carl Thunberg (1743–1828).

CAROLI PETRI THVNBERG
MED. DOCT. PROF. REG. ET EXTRAORD. ACADEM.
CAES. N. C. REG. SCIENT. HOLMENS. SOCIET. LITTER.
VPSAL. PATRIOT. HOLMENS. BEROLIN. N. SCRVT.
LVNDIN. HARLEMENS. AMSTELDAM. NIDRO-
SIENS. MEMBRI

FLORA
IAPONICA
SISTENS
PLANTAS
INSVLARVM IAPONICARVM
SECVNDVM
SYSTEMA SEXVALE EMENDATVM
REDACTAS
AD
XX CLASSES, ORDINES, GENERA
ET SPECIES
CVM
DIFFERENTIIS SPECIFICIS, SYNONYMIS PAVCIS,
DESCRIPTIONIBVS CONCINNIS ET
XXXIX ICONIBVS ADIECTIS.

LIPSIAE
IN BIBLIOPOLIO I. G. MVLLERIANO
1784.

Title page of Thunberg's *Flora Japonica*, 1784.

Viburnum opulus.

Thunberg's '*Viburnum*' (later *Hydrangea*) *macrophyllum*.

plants *Viburnum macrophyllum* and *Viburnum serratum*, the latter with smaller leaves and more slender twigs. His mistaken identification was because of their striking similarity in flower to the European *Viburnum opulus*, and perhaps also to *Viburnum plicatum*, which he might also have seen in Japan. The genera have two things in common: a flat corymb of small, 'real' or fertile flowers surrounded by a ring of broad-petalled ray flowers, and forms with a globose head crowded with the showy ray flowers.

At the same time, a Chinese botanist introduced a Chinese garden variety of hydrangea to France. Philibert Commerson, a French botanist, was said to have named this plant *Hortensia* to commemorate a lady named Hortense, whose identity has never been definitively established. Jussieu, in his widely used 1789 *Genera Plantarum*, published this plant under the name *Hortensia*, the same year as Joseph Banks introduced his Chinese garden-originated plant to Kew; and when, in 1830, the French botanist Seringe transferred Thunberg's two goat-fodder 'viburnums' to the genus *Hydrangea*, the two species names Thunberg had chosen when describing them as viburnums were preserved in the names *Hydrangea macrophylla* and *Hydrangea serrata*. (Both are now classified in the botanical subsection *Macrophyllae*: see Chapter 6). The names *Hortensia*

Thunberg's *'Viburnum'* (later *Hydrangea) serratum*.

and *Macrophylla* are used interchangeably to describe cultivars of *H. macrophylla*.

A few decades later, in the mid-nineteenth century, another employee of the Dutch East India Company, Philipp von Siebold, a German physician and eye specialist, was also consequential in bringing more Japanese hydrangeas to Europe. Because of his specialist skill as an ophthalmologist, the authorities allowed him to live – and therefore be able to botanise – on the Japanese mainland. He introduced many other Japanese plants, some now well-known and dedicated to his name, such as *Magnolia sieboldii*, *Malus sieboldii*, *Hosta sieboldii* and *Thuja sieboldii*. Among the hydrangeas he introduced was *Hydrangea paniculata*, which is today, along with *Hydrangea macrophylla*, one of the most widely planted species.

In 1862, von Siebold also introduced **'Otaksa'**, a seminal cultivar of *H. macrophylla* in the history of the development of hortensias, quite different morphologically from the original Banks' introduction, with smaller leaves, thinner shoots, more compact habit and less vigour. The name derives from Otaki-san, the reportedly beautiful Japanese 'wife' of von Siebold.

Wild form of *H. paniculata* with pink buds (bottom). Its cultivars are now among the most widely planted hydrangea in gardens. Same wild form close up (top).

Discovered carrying illegal maps of Japan, von Siebold was banished by the Japanese Government and returned to Europe. But he was determined to return, and was later able to go back to Japan, where he is now one of the best-known and respected historical botanists from the West. He then fully immersed himself

Siebold's Japanese 'wife'.

Hydrangea macrophylla 'Otaksa', named after Siebold's Japanese 'wife'.

in Japanese culture, setting up a school teaching both medicine and botany – at the time closely linked disciplines.

A Russian botanist, Carl Maximowicz, who was allowed some access to the flora in Japan in 1860, also collected some hydrangeas there, among other plants: several of these, hybrids and wild forms of both *H. macrophylla* and *H. serrata*, eventually arrived in Europe.

Hydrangea macrophylla in paintings.

The origins of the seminal cultivars

However, the basis for the development of what has become the modern *H. macrophylla* industry was really only firmly established with the introduction of two further cultivars by Charles Maries, a horticulturalist employed by James Veitch to collect plants in Japan. Veitch was the biggest and most influential nursery firm in England in the late-nineteenth and early-twentieth centuries, and sponsored collectors to seek and collect plants all over the world. In 1879, Maries sent back *H. macrophylla* **'Mariesii'** and *H. macrophylla* **'Rosea'**, both in fact hybrids of indeterminate origin, but with the limited knowledge of the genus at the time, regarded then as distinct species.

In 1881, a 'lacecap' type, in the floral style of the wild species, *H. macrophylla* **'Veitchii'**, which Charles Maries is said to have brought in from Japan, was added to the mix (although this is now disputed as a Maries' introduction). These were the plants generally available for breeding towards the end of the nineteenth century that enabled the emergence of *Hydrangea macrophylla* as a decorative pot plant.

Hydrangea macrophylla 'Mariesii', halfway between a lacecap and a mophead cultivar.

Hydrangea macrophylla 'Rosea', often seen as an attractive blue: naming hydrangeas after their colours is always a risky business.

Hydrangea macrophylla 'Vietchii', still highly valued in gardens.

French initiative: potted plants of the first rank

In 1903, French nurseryman Victor Lemoine sowed seeds of *H. mariesii*, which yielded plants of very different character from the parent, showing it to be of hybrid origin. He named

Hydrangea macrophylla 'Mariesii Perfecta'.

Hydrangea macrophylla 'Mariesii Grandiflora'.

Hydrangea macrophylla 'Mariesii Lilacina'.

Hydrangea macrophylla 'Mme Emile Mouillere' (left) and *H. macrophylla* 'Générale Vicomtesse de Vibraye' (right). Two seminal cultivars still popular today.

Hydrangea macrophylla 'Merveille Sanguine' with a reversion back to the original pink *H. macrophylla* 'Merveille'.

three, *Mariesii perfecta*, *M. grandiflora* and *M. lilacina*, and remarkably all are still available and widely grown today, 120 years later.

Lemoine also spotted that among the showy sterile florets of plants with globose flower heads, there were a few fertile flowers. These yielded seeds, which he raised, and which in 1908 enabled him to launch a series of notable globose-flowered mophead novelties. At the same time, also in France, fellow nurseryman Emile Mouillere crossed these new seedlings with *H. macrophylla* 'Rosea' and introduced a whole new range, including two of the best mophead varieties for outdoor use today: the big white *H. macrophylla* 'Madame Emile Mouillere' and the butterfly blue *H. macrophylla* 'Générale Vicomtesse de Vibraye'.

Henri Cayeux, another French breeder, introduced *inter alia* another fine variety, *H. macrophylla* 'Merveille', a deep-pink mophead, from which a branch sport known as 'Merveille Sanguine' arose, the darkest red mophead I have seen in cultivation.

Following the initiatives of Lemoine and Mouillere the scene was set for others to produce more varieties in France, Germany, Holland and Switzerland. The focus was entirely on producing colourful globose-headed flowers derived from *H. macrophylla*. These were selected for pot plant and conservatory use, rather than as garden plants, and many hydrangeas with a lacecap format that would today be considered as first-class garden plants were discarded in favour of these large-flowered globose mopheads. Like remontancy now in the rose industry, with breeders discarding many fine roses because they are not remontant but flower only once, this historic narrowing of selection still influences public appreciation of the genus *Hydrangea* today. The French found, too, that hydrangeas were easy to force into flower, which extended the flowering time-frame for pot plants, and thus expanded sales for that particular form of the genus.

British indifference

It remains something of a mystery why, as France became the main centre for this expanding and profitable activity in the early twentieth century, the neighbouring British played little or no part in it. It appears that in Britain, hydrangeas were simply not a horticultural priority in the early part of the twentieth century, and no British nursery appeared to have the foresight to capitalise on new developments and promote hydrangeas for profit, as happened in France. It may be that the market for pot-grown plants in Britain was not as developed as in Europe. In the case of hydrangeas, in the cold winters then prevailing on

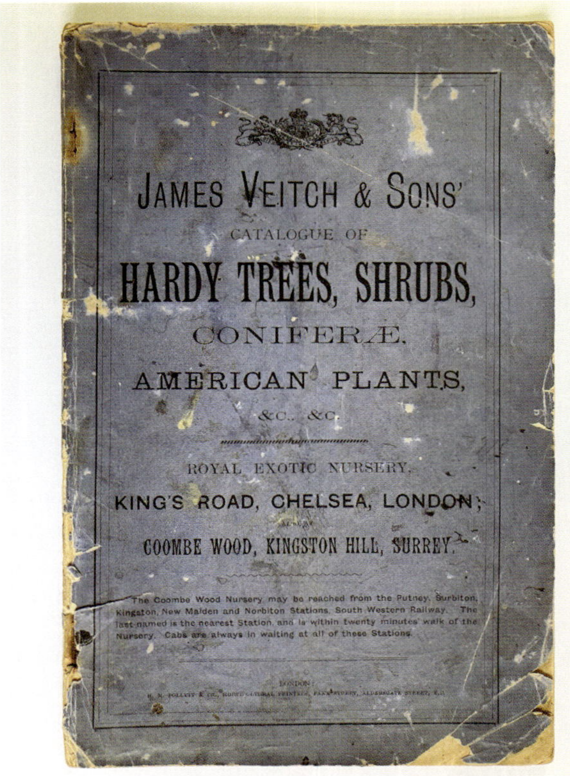
JAMES VEITCH & SONS'
CATALOGUE OF
HARDY TREES, SHRUBS,
CONIFERÆ,
AMERICAN PLANTS,
&C., &C.

ROYAL EXOTIC NURSERY,
KING'S ROAD, CHELSEA, LONDON;
AND AT
COOMBE WOOD, KINGSTON HILL, SURREY.

The Coombe Wood Nursery may be reached from the Putney, Surbiton, Kingston, New Malden and Norbiton Stations, South Western Railway. The last-named is the nearest Station, and is within twenty minutes' walk of the Nursery. Cabs are always in waiting at all of these Stations.

LONDON:

A rare nineteenth-century Veitch catalogue, courtesy of Caradoc Doy.

H. macrophylla 'King George', one of the Chelsea Gold Award-winning hydrangeas of H. J. Jones in the mid-1920s.

the Continent, the principal way of cultivating them there successfully was in pots, so they could be easily winter protected. Their garden use was of secondary importance.

Veitch, a major British force in horticulture at the turn of the century for other genera, strangely seems to have made little use of the new hydrangea plants introduced by Maries or the early European hybrid introductions; and even with the eventual advance of the hydrangea as a garden plant, Britain has continued to be something of an exception in this history of development in the genus, despite having both a climate and existing gardens with ideal conditions and opportunities to exploit its full range.

Not many, for example, have heard of British hydrangea breeder H. J. Jones, who was awarded gold medals for his hydrangea exhibits at Chelsea in the mid-1920s. Only one or two of his cultivars are in cultivation in Britain today, more perhaps to be found in France, and his name has disappeared into the hydrangea mists of time (*see* Chapter 6).

It is worth considering the possible causes of this relative indifference in the UK to a superb garden plant that provides showy colour exactly when other staples of English spring and early summer gardens are over. Different factors come to bear at different times.

Crucially, hydrangeas in Britain were probably not considered as suitable garden plants for outdoor use because they were initially assumed not to be hardy. The same assumption at first applied to other exotica from Asia, such as camellias. Both genera were initially simply not trusted outside in the unpredictable English climate, and were cosseted in greenhouses or conservatories as stove plants. Both were also seen as vulnerable to frost damage beyond repair: something only relatively recently understood as false.

This misperception about lack of hardiness was probably confirmed in the public mind, for example, when soft mophead hydrangeas, straight from the nursery greenhouse, forced into early flower and used as houseplant gifts for Christmas and winter birthdays, were planted in the garden in late winter or spring, after flowering, straight from a heated house. Their soft shoots were either immediately frozen or consumed by slugs, leaving little trace. This firmly reinforced the view that hydrangeas in damp British gardens were inevitably martyrs, if not to frost, then to slugs and snails. They also lacked fragrance.

Perhaps some influential horticulturalists also regarded the rather overblown flowers of the mophead forms as coarse and vulgar – a phenomenon precisely of continental fashion, thus out of place in the great individual gardens of the British shires, which prided themselves on being showcases for botanical collections and newly created garden hybrids in a natural, informal landscape. British gardens favoured plants grown for total landscape effect, where rhododendrons, camellias and magnolias took front and centre stage. They perhaps saw the mophead forms of *H. macrophylla*, then known mostly as pot plants with showy 'artificial' flowers, out of context in this setting.

As Christopher Lloyd wrote with typical impish humour in *The Well-Tempered Garden* in 2001:

'The mopheaded hortensias… are regarded, in refined circles, as crude, blatant, obvious, coarse, vulgar. In that case I must have something of all those qualities myself. I do not like these hortensias at all times and in all places, but they have a tremendous luxuriance and vitality that one cannot help admiring. The wild types of hydrangea are known as lacecaps… but these are the only kinds worth growing according to the sensitive (but selfconscious) man of taste'.

Vive la différence: lacecap (left) versus mophead (right) *Hydrangea* macrophylla forms.

Indifference among the British

Some 45 specialist plant societies in the UK are devoted to individual genera, such as carnations, dahlias, clematis and so on, reflecting a dedicated gardening nation. However, it is remarkable that even today, although there are hundreds of hydrangea species and cultivars listed as commercially available in the RHS Plant Finder, and every garden centre in the country sells a range of hydrangeas, no British plant society exists for hydrangeas.

The first European conference on hydrangeas was held in 2007 in Belgium, and it attracted more than 200 delegates from all over the world, many coming in numbers from as far afield as New Zealand, Japan and America. There were just four paid-up delegates from the UK (excluding lecturers and RHS exhibitors presenting trial results).

John Massey's Ashwood Nursery's hydrangea stand at RHS Chelsea, 2005.

John Massey and Philip Baulk of Ashwood Nurseries memorably put on a magnificent Gold Medal display at the Chelsea Show in 2005 of more than 80 hydrangea cultivars, expertly brought forward in flower and superbly displayed in a variety of colour and form, and by any objective measure a contender for the award of best stand in the show. I remember looking on in some amazement as the BBC TV team walked straight past it, to a stand featuring grasses. In a whole week of nightly TV programmes it received scant, if any, coverage, suggesting that the producers took the view that this outstanding exhibit of hydrangeas was of minimal public interest.

Hydrangea aspera 'Hot Chocolate' (left) and *H. aspera* 'Rosemary Foster' (right).

It may be that the Victorians whimsically saw the hydrangea in a negative light because of the then fashion for floral symbolism, equating it with vulgarity and vanity, a big showy flower lacking in real substance, with its colour changes a sign of unreliability, or change of heart. For example, when sent as flowers by men to women who had turned down their advances, the pale blue hydrangea was symbolically read as an emblem of the woman's coldness.

Perhaps, too, the rather elongated foreign names, often not descriptive but celebrating European persons of little note in UK social circles, did not trip easily off an English tongue. What the French see as English parochialism could have played a part: a British inclination to mistrust any language other than English; to turn an English perception of the French on its head, '*l'exception anglaise*'.

There are probably other examples of a yawning (in both senses) British indifference to the genus. But there are signs that its garden value is being increasingly recognised, and a more informed appreciation of the hydrangea is steadily growing. The British gardening public is waking up to its merits.

This is partly the effect of a recent move to recognise other species and cultivars as having garden value as well as *H. macrophylla*, hitherto the only popularly known form of the genus. Having said that, over the last 50 years I have found it difficult to interest nurserymen in the UK in new varieties that are not *H. macrophylla* cultivars – in contrast to the interest shown in new *H. serrata* and *H. aspera* seedlings and hybrids (such as *H. aspera* 'Hot Chocolate' and *H. aspera* 'Rosemary Foster') by European nurserymen.

Until recently, although new cultivars of *H. aspera* and *H. serrata* demonstrate huge improvements in flower and foliage on what has gone before, present as first-class hardy flowering shrubs and have trialled successfully outdoors at White House Farm in Kent for years, their novelty value has appeared to count for little in the UK's mass horticultural market.

Yet in Europe a wide range of hydrangeas beyond *H. macrophylla* are already valued highly as flowering shrubs for garden planting. In America, there have been competitive breeding programmes and an explosion of new varieties, with the hydrangea there being called the plant of the Millennium, characterised by vigorous commercial promotion that is conspicuously lacking in the UK. The present rapid expansion of the genus in the United States is arguably one of the more significant events currently happening in horticulture in that country. While the focus in the USA is still mostly on *H. macrophylla* cultivars, thanks *inter alia* to the promotion of 'remontancy' (*see* Chapter 6), other species are now also being brought into play.

Haworth-Booth and the *Hydrangea* as a popular garden plant

There is one exception to this history of relative British indifference: the British horticulturalist Michael Haworth-Booth. In the mid-twentieth century, his writings and exhibits at RHS Vincent Square flower shows in London brought hydrangeas to the attention of a much wider audience. His book *The Hydrangeas*, the first dedicated title to be published on the subject in the UK, ran to five reprints: the first edition published in 1950, a further edition becoming a Garden Book Club selection and the last, fully revised, published as late as 1983.

This was essentially a horticultural work, instrumental in focusing on the hydrangea's value as a plant for the average garden, rather than merely as a subject for pots – a crucial issue. It was not until after the war in the 1950s that the real value of

Haworth-Booth's seminal books of 1950, 1976 and 1983, introducing the hydrangea to a wider British public.

the hydrangea as a garden plant really began to be appreciated in the UK, and this was due in no small part to Haworth-Booth's knowledge and ability to articulate how to get the best out of them, in gardens large and small. Nonetheless, like all books published in the UK on the subject, the focus of his book was primarily on the Hortensia-style macrophyllas.

My own interest in hydrangeas began when I worked briefly for Haworth-Booth as a casual labourer in the mid-1950s during the university long vacation. I have many of his plants in my collection and parts of my now mature garden model the effective long-term results of his planting methods. My interest in the genus picked up pace significantly in the 1990s, when I was finally able to expand beyond the Hortensia varieties and began collecting seed of wild hydrangea species in China in the 1990s, later visiting Japan, where *H. serrata* in particular is nationally treasured.

Three Teller cultivars: 'Moewe' (seagull) (left and top); 'Rotschwanz' (redstart) (centre) and 'Pfau' (peacock) (right) in neutral soil.

From Hortensia lacecaps in the 1970s to paniculatas today

The so-called 'lacecap' varieties – Haworth-Booth's coinage for what the Swiss would later call the more prosaic German *Teller* ('plate') – have emerged as a significant feature of the hydrangea story in the late-twentieth century. This is a term for those flowers closer to the style of the wild species, with a flat or gently convex ('cap'-shaped) rather than globose corymb, consisting of a disc of numerous tiny fertile flowers surrounded by a peripheral ring of relatively large and showy sterile 'ray' flowers. Typical of these, and among the first to be deliberately developed, are the Swiss-raised Teller Series of 26 cultivars created, marketed and named after birds and distributed for propagation during the latter half of the last century (discussed in Chapter 6).

Initially this series was bred specifically for the pot-plant market as an alternative choice to the traditional globose mophead forms. But collections were eventually sent to large well-known gardens across Europe for garden trial, and most Teller cultivars subsequently turned out to be excellent hardy garden plants. Marketing a 'series', where crosses have yielded a range of similar saleable plants, is a modern phenomenon that is essentially a marketing tool, but has some value to the gardener seeking to make planting choices.

Growing the future

There is undoubtedly growing interest at present in the mass and range of colours that hydrangeas can bring to the summer garden – of which this book is evidence. Many big gardens open to the public that rely on the dramatic spring display of rhododendrons, camellias, cherries, lilacs and magnolias and so on, can become something of a green desert after Easter, except for formal rose beds and labour-intensive and expensive herbaceous borders. These gardens need summer footfall, and some are now recognising that hydrangeas can create a drawcard colour factor to bring in the paying customer, extend their season by a quantum amount, and are cheap and easy to maintain: a prominent and relatively trouble-free source of colour and interest.

Alongside the colourful hortensias, in recent years there have also been significant advances in developing new cultivars of both *H. paniculata* and *H. arborescens*, totally hardy and thus suitable for cold gardens anywhere, of any size. In particular, there has been a recent explosion in the numbers of *H. paniculata* introduced to the market, with a potentially bewildering range of similar choices for the gardener (I understand there are now some 140 named cultivars of *H. paniculata* in the UK national collection at Derby). *Hydrangea arborescens* has produced 'Annabelle', perhaps the most widely planted hydrangea in today's gardens.

Other new species, such as *H. chinensis* and *H. scandens*, are now being used to make new hybrids that extend the range of flower styles and growth habit: for example, the excellent white free-flowering *H.* 'Runaway Bride', plant of the year at Chelsea Flower Show 2018.

Hydrangea 'Annabelle', a mophead form of *H. arborescens.*

Over the last 30 years, *H. serrata* in its many forms has also gradually found its way into Western gardens. It is even more varied in form than its cousin *H. macrophylla*, but always smaller in stature and more understated and elegant in its appeal. The arrival of these hydrangeas as garden plants, very different to the hortensias, occurred mostly after World War II, and has accelerated in recent years; this is treated in further detail in Chapter 6.

At the moment the focus of marketing worldwide is still principally on the multicoloured hortensias (now being joined, to some extent, by the explosion of *H. paniculata* cultivars). This book aims to enlarge that picture: in particular, to show how versatile other types of hydrangea are as first-class flowering shrubs for the modern garden.

Hydrangea 'Runaway Bride', Plant of the Year at Chelsea 2018 – *as a H. scandens* a whisper of change?

Wild form of *H. serrata* in Japan, a species highly valued by the Japanese.

CHAPTER 2

DISTRIBUTION AND CLASSIFICATION OF SPECIES

The Geographical Divide

The two earliest introductions, *H. arborescens* from America and *H. macrophylla* from Japan, are representative of different species in what are two wholly disjunct, quite separate populations in the wild. The genus divides between the temperate areas of East Asia and eastern North America, with related species linking both areas.

This geographical division of related plant populations between America and Asia is not at all unusual. There are an estimated 65 similarly-related genera split between these two continents – *Crataegus*, *Magnolia*, *Lindera* or *Spiraea*, for example – reflecting the past close relationship between two now geographically separate floras. Elements of Hydrangeaceae and many other genera probably migrated overland from Eurasia when North America was contiguous with Europe and Asia; and survived and developed independently on the basis of their ability to adapt to changing climatic and geological conditions.

In addition to these two main areas of hydrangeas, there are also the evergreen liana-style hydrangeas that climb trees and cliffs in the subtropical and warm temperate climate of Central and South America. But at the moment these are of minor significance in our story, as with one or two exceptions, the majority are not hardy and thus not suitable for cultivation in cool temperate gardens. I say 'at the moment' as some of these plants are of great floral potential and could play a significant

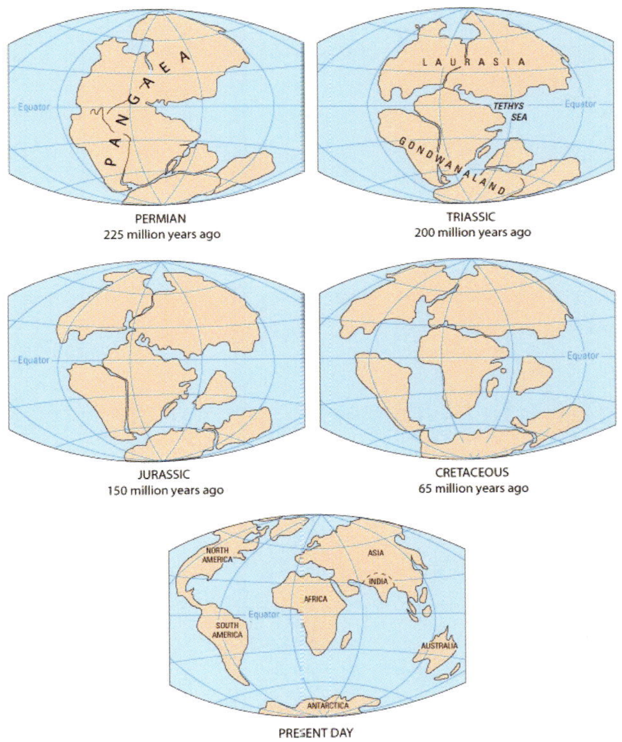

Map showing the Asian and American continents before and after separation.

OPPOSITE PAGE: *Hydrangea aspera* 'Rosemary Foster', a White House Farm hybrid.

part in the development of future climbing hybrids of acceptable hardiness (*see* Chapter 13).

The American hydrangea species are confined to the mountains in the east and south of the United States, mainly in the Appalachians, but the area of Asian hydrangea species' distribution extends over a much wider area, along the Himalaya from Nepal to western, central and southern China, through to Japan, Korea, Taiwan, Sumatra and Java. In addition to the vitally important Japanese species, China has lately and unsurprisingly been recognised as a centre for all kinds of recently introduced hydrangea species and forms, particularly in its western provinces, where the influence of the monsoon and a happy combination of latitude and altitude provide an unrivalled favourable climate that sustains some of the richest and most diverse temperate woody plant populations in the world. Recent introductions of hydrangea species and forms since the early 1980s from China are likely to have an influence on further hybrid developments, so we are certain to see significant additions to the range of interesting and gardenworthy cultivars in the future.

But today, it is the variety and colour of Japanese hydrangeas spreading from this turn-of-century first flourish of interest that make up the vast majority of the plants in gardens everywhere. While there are many excellent and, I would argue, underappreciated and underexploited species in China, it is the three Japanese species, *H. macrophylla*, *H. paniculata* and *H. serrata*, that make up the bulk of popular cultivars grown in gardens across the world. Their colour, form and flower power make them the cornerstone of the substantial hydrangea nursery trade in Europe and, in recent years, the USA.

Classification and taxonomy: the McClintock monograph

Most horticulturalists agree that the most significant occurrence of the genus in the wild – in terms of both scale and horticultural potential – is in East Asia. But total numbers of species vary widely in the literature, depending on which authority you follow taxonomically. Elisabeth McClintock wrote the first comprehensive monograph of the genus in 1957, when she was Assistant Curator of Botany at the California Academy of Sciences. This has remained the most important work of reference since then, despite a degree of qualification by subsequent authorities. Her revision in 1957 was certainly needed, as previously many plants had been named as separate entities by a raft of individual explorers and botanists from different countries. This is perhaps unsurprising, as throughout their range, hydrangea species' polymorphism – a multiplicity of forms within a species – is prevalent, so plants with relatively minor differences were often identified as separate individual species.

Floral variability within subsection Asperae: twelve forms of *H. aspera* cut from my garden, no two quite the same.

Hydrangeas fall into two main sections – Hydrangea and Cornidia. These are further subdivided into subsections, each with its own morphological character.

McClintock examined no less than 150 different taxa during her study, and was able to radically reduce the number of species to 23. She set this out in a letter to me: 'You mentioned having seen or used my monograph; perhaps you noticed that I have only about 20 species in the genus and not the 60 or 70 listed by Haworth-Booth and others who have written about hydrangeas. In making a careful examination of the entire genus it is my opinion that the number of species in *Hydrangea* has been greatly exaggerated, and most of them I have listed in synonymy.' She subsequently visited White House Farm and I enjoyed the great privilege of talking hydrangeas with her.

More recent revisions

For the most part, McClintock's 'lumping' of species (the inclusion of many species under a single umbrella classification) has been regarded as justified, but in some cases it has been felt to be an oversimplification. For example, Bean, in his authoritative *Trees and Shrubs Hardy in the British Isles* (last edition 1973) was rather equivocal in his reaction to McClintock, admitting that while following her in lumping many familiar species together, he maintained others at species level, but only, as he put it, where there was 'taxonomically some room for manoeuvre'. In other words, he accepted that a flexible approach was permissible if a reasonable case could be made, though admitting it was more for the convenience of gardeners than from any conviction that McClintock was wrong.

By contrast, in their *Flora of China*, Wei and Bartholomew (2001) characterise some 73 species in total, more than three times as many as McClintock, with 33 in China, of which 25 are endemic. But in the same work, the two authors failed to reach agreement, Bartholomew arguing that it should really be reduced to almost half that number: eighteen species in China, with nine endemic. This is typical of the conflict in taxonomic opinion that continues among botanists today.

The McClintock revision

Section 1: Hydrangea		
Subsection 1: **Americanae** (2 species)	1 *Hydrangea arborescens*	*H. arborescens* ssp. *arborescens*
	2 *Hydrangea quercifolia*	*H. arborescens* ssp. *discolor*
		H. arborescens ssp. *radiata*
Subsection 2: **Asperae** (3 species)	1 *Hydrangea sikokiana*	
	2 *Hydrangea involucrata*	*H. involucrata* ssp. *longifolia*
	3 *Hydrangea aspera*	*H. aspera* ssp. *aspera*
		H. aspera ssp. *strigose*
		H. aspera ssp. *robusta*
		H. aspera ssp. *sargentiana*
Subsection 3: **Calyptranthae** (1 species)	1 *Hydrangea anomola*	*H. anomola* ssp. *anomola*
		H. anomola ssp. *petiolaris*
Subsection 4: **Petalanthe** (2 species)	1 *Hydrangea hirta*	
	2 *Hydrangea scandens*	*H. scandens* ssp. *scandens*
		H. scandens ssp. *liukiuensis*
		H. scandens ssp. *chinensis*
		H. scandens ssp. *kwangtungensis*
Subsection 5: **Heteromallae** (2 species)	1 *Hydrangea paniculata*	
	2 *Hydrangea heteromalla*	
Subsection 6: **Macrophyllae** (1 species)	1 *Hydrangea macrophylla*	*H. macrophylla* ssp. *macrophylla*
		H. macrophylla ssp. *serrata*
		H. macrophylla ssp. *stylosa*
		H. macrophylla ssp. *chungii*
Section 2: Cornidia		
Subsection 1: **Monosegia** (8 species)	*Hydrangea seemannii*	
	Hydrangea asterolasia	
	Hydrangea integrifolia	
	Hydrangea oerstedii	
	Hydrangea peruviana	
	Hydrangea diplastemona	
	Hydrangea preslii	
	Hydrangea steyermarkii	
Subsection 2: **Polysegia** (4 species)	*Hydrangea serratifolia*	
	Hydrangea farapotensis	
	Hydrangea jelskii	
	Hydrangea mathewsii	

In 2015 and 2016, different taxonomic revisions of the genus have continued this tendency towards 'lumping' rather than 'splitting', with more recent proposed changes based solely on interpretation of molecular phylogenetics in the laboratory. This has radically shuffled the pack of species, coming up with relationships and classifications that now often leave out of account the morphology, habitats and ecological preferences of the plants.

That this laboratory revision work pays little heed to the needs of gardeners is hinted at by Bean's comment about 'taxonomic room for manoeuvre'. Its use value is unfortunately rarely questioned. I have argued elsewhere there should be a better balance between perceived botanical accuracy and empirical knowledge in practice in the field, in the effort to achieve long-term clarity and taxonomic security. Morphology – the science of form, analysing the structural physical characteristics of plants – was always the traditional method of establishing plant relationships, and is the basis of the gardener's practical empirical plant knowledge and perceptions over generations. Botany itself began by seeking to discover close relationships between plants through morphological analysis and comparison, in order to promote a better understanding of their character and processes; but

'New' Hydrangeas

The following previously designated as separate and well-defined genera within the family Hydrangeaceae, are now 'sunk' into the genus *Hydrangea* by Samain and de Smet following phylogenetic laboratory analysis. They are now hydrangeas, with proposed name changes to reflect this.

Broussaisia	A herbaceous perennial endemic to Hawai'i. Also known as kanawao in the vernacular.
Cardiandra	Five species of herbaceous perennials and subshrubs. Not often seen as they need both shade and warmth to thrive in the garden.

Broussaisia.

Deinanthe.

Schizophragma.

Cardiandra.

Dichroa.

Decumaria.

Platycrater arguta.

Pileostegia.

Decumaria	Two species of climbers, *D. barbara* in the USA and *D. sinensis* in China. Both have white inflorescences that consist only of small fertile flowers. The type is now called *Hydrangea barbara*.
Deinanthe	Two species of herbaceous perennials for the woodland: *D. bifida* from Japan and *D. coerulea* from China. The name derives from the Greek '*deinos*' (strange) and '*anthos*' (flower). Now called *Hydrangea deinanthe*.
Dichroa	About a dozen species, the best known of which is *D. febrifuga*, with blue flowers and fruits (now called *Hydrangea febrifuga*). Few are reliably hardy, but crosses with *Hydrangea macrophylla* have yielded hardier lookalikes with blue flowers (*Didrangea – see* Chapter 13).
Pileostegia	Photographed in China in the wild. Four species of this self-clinging evergreen climber, of which *P. viburnoides* is usually seen in cultivation. Flowers cream/white in terminal panicles. Shade tolerant on walls and into trees. Now called *Hydrangea viburnoides*.
Platycrater arguta	A low deciduous shrub to about 1m (3ft) that enjoys a cool position in light shade. Flowers are small and white with yellow stamens in a loose terminal cyme; the ray flowers are small 2cm (1in) flimsy discs of pale green. Now called *Hydrangea platyarguta*.
Schizophragma	A genus of some ten species of self-clinging climbers, two of which are frequently seen in gardens: *S. hydrangeoides* from Japan, with two named cultivars, 'Roseum', with rose pink bracts, and 'Moonlight', with silvered, bluish foliage (both have been awarded the RHS AGM and are now called *Hydrangea hydrangeoides*); the second is *S. integrifolium*, a climber to 15m (45ft) also with an AGM, with floral bracts some 6–8cm (2–3in) long, forming large multi-flowered inflorescences.

The above, including the herbaceous perennials, are all now designated as hydrangeas (Samain and de Smet 2015).

Marie-Stephanie Samain defended this radical shift while admitting 'Such a change may understandably cause consternation among those not familiar with molecular phylogenetic studies…' (*The Plant Review*, December 2021).

A further interpretation by Ohba and Hakiyama in 2016 sought to rescue some of these sunk genera with a further revision of this revision, reclassifying yet again many of the sections and species, and reminding us that these molecular methods are themselves not definitive, but subject to rival interpretation among botanists.

recently, DNA analysis has often been accepted unreflectively as gospel, as a kind of incontrovertible and definitive evidence of evolutionary relationships. This overturns many old accepted links, used and understood by generations of gardeners

For example, in a 2015 analysis, botanists Samain and de Smet transferred into the genus *Hydrangea* eight genera within the family Hydrangeaceae that could not be more different or distinct in morphological terms. This radical move means a loss of definition and clarity for the gardener, and a consequent reduction in the use value of the name '*Hydrangea*'. It renders *Hydrangea* taxonomy a mile wide and an inch deep.

The taxonomic situation is thus confused at present, not to say anarchic, certainly for horticulturists, if not for evolutionary botanists. There is no law that every new laboratory diktat must be followed, and as McClintock is the major reference dealing with the genus as a whole, in this book I shall follow her taxonomy as my base for discussion and amendment; while also adopting the Bean approach of pick 'n' mix as best serving the interests of horticulturists keen to strike a balance between botanical accuracy and the need for empirical horticultural clarity.

In this context it is worth saying that this book is essentially for practical horticulturalists, landscapers and designers eager to know more about hydrangeas, and who may perhaps find some stimulus to plant them and get value from them in their gardens. While taxonomic shenanigans may feed the curiosity and interest of the investigative gardener, they are often of little practical horticultural help.

CHAPTER 3

AMERICANAE: THE 'ANNABELLE' PHENOMENON

Driving south from Raleigh in North Carolina with the late Dr Clifford Parks, a friend and an expert on the local flora there as well as a world authority on the genus *Camellia*, we explored the Blue Ridge Parkway with its extensive forest views, and hiked mountain trails into the Appalachians amid unspoiled forests of oak, liquidambar, lime, beech, birch, maple, pine and spruce. Its understorey included native wild rhododendron, magnolia, kalmia, fragrant deciduous azaleas – and hydrangeas.

Cliff was an untiring source of information about this rich and diverse flora, notable as home to the very first ever hydrangea to be introduced and cultivated in England – *Hydrangea arborescens*, one of the two species in the subsection named, for obvious reasons, Americanae. The other species in the subsection Americanae is *H. quercifolia*. Both originate in the eastern and southern USA, but *H. arborescens* is found in the wild from New York state down to Louisiana, while *H. quercifolia* s more confined to the southern states. It is the state flower of Alabama.

Clifford Parks with species-rich forest in North Carolina (Looking Glass Mountain).

North Carolina forest, home to *Hydrangea arborescens.*

Arborescens distribution and variants

Hydrangea arborescens was first introduced into England as early as 1736. It found its way into Europe soon afterwards, but

OPPOSITE PAGE: 'Sweet Annabelle' rivals the bestselling white form for vigour and tolerance of tough conditions and is one of the best to emerge from the scramble for a good pink.

was never ornamental enough to become popular. Not the most colourful of species, but with some novelty value, as Philip Miller put it in 1768, '...it is now preserved in gardens for the sake of variety more than its beauty'. That situation was to change radically in the late-twentieth century, as the most famous of its anomalous forms, *H. arborescens* **'Annabelle'** was named and marketed: now perhaps the most widely planted hydrangea cultivar in gardens over the past decade.

Hydrangea arborescens and its subspecies are widespread in the eastern USA, stretching down the Appalachian mountain chain through the Carolinas into the southern states of Louisiana, Mississippi, Kentucky, west into Indiana and as far north as southern New York state, through Maryland, New Jersey and Virginia.

The *arborescens* name is a misnomer, as typically the plant is not ambitiously tree-like, but a rather loose and straggly shrub, up to 1–1.5m (3–5ft) in cultivation, though some specimens can put on more height. The typical inflorescence of *H. arborescens* as a wild species is flat or domed, sometimes with small ray flowers scattered around the periphery. I say 'sometimes', as frequently these ray flowers are absent altogether, leaving a grey/white rounded dome of only tiny, densely packed fertile flowers, of no great beauty.

The type species *H. arborescens* ssp. *arborescens* is known as the 'smooth hydrangea', with a glabrous or hairless leaf underside. Leaves are ovate or elliptic, from 5–15cm (2–6in) long and about twice as long as wide. In addition to this type species, there are two subspecies, differing principally in their leaf vestiture, with more or less pubescence (hairiness) on the leaf reverse. Of the two more pubescent or hairy subspecies, *H. arborescens* ssp. *discolor* has a lower leaf surface that is more or less tomentose, the hairs on the reverse giving a greyish effect. *H. arborescens* ssp. *radiata* in contrast has a close, dense indumentum, giving the leaf reverse a conspicuous silvery/white effect.

For all the Americanae, the absence or presence of sterile ray flowers varies, depending not only on the subspecies but also on the geographical area in which it grows. McClintock reports

Examples of varieties of *H. arborescens*, which can occur with and without ray flowers.

The three subspecies' leaf reverse, showing varied degrees of vestiture: (left) ssp. *arborescens*, (centre) ssp. *discolor* and (right) ssp. *radiata*.

Hydrangea arborescens ssp. *radiata,* the most gardenworthy wild form.

Hydrangea arborescens ssp. *radiata* 'Samantha', a mophead form of the species, which I have found to have a weak constitution.

Not all *H. arborescens* ssp. *radiata* in the wild have copious and conspicuous ray flowers; a helicopter form shows the variability that makes up the corymb.

Hydrangea arborescens 'Annabelle'.

that in the eastern part of its range from Virginia to South Carolina, of 231 individuals examined, only 65 (27 per cent) had ray flowers. In the western part of its range, from Tennessee and Kentucky to Missouri and Arkansas, of 243 individuals examined, 167 (69 per cent) had ray flowers. The presence or absence of sterile ray flowers is somewhat associated with the degree of pubescence on the reverse of the leaf. This was evident, for example, with *H. arborescens* ssp. *discolor*, which in the eastern part of its range scored only 17 per cent in its ray flower count, while in the western area, some 92 per cent had sterile ray flowers.

Hydrangea arborescens ssp. *radiata*, with the densest indumentum on the leaf reverse that creates a very attractive silvery/white effect, scored almost 100 per cent for ray flowers, some more spectacular than others. So from a gardening perspective, ssp. *radiata*, for its combination of both indumentum and ray flowers, is the most appealing form of the seminal wild species.

Arborescens cultivars: 'Annabelle' and its relatives

From this horticulturally inauspicious base, and the relative lack of ornament of the wild species, a breakthrough for *H. arborescens* as a desirable garden plant came relatively recently, with the rediscovery and commercial introduction of the big sterile mophead form **'Annabelle'**, making *H. arborescens* since then the source of some of the most gardenworthy hydrangeas for general planting.

Like the Heteromallae section of the genus, the mountainside hardiness of *H. arborescens*, its extraordinary tolerance of cold and heat, and its suitability for any soil, make this a plant for anywhere, in any garden, sun or shade. During the record-breaking heatwave of 2022 in Europe, where for several weeks temperatures were in the mid-thirties, in some places reaching over 40°C, *H. macrophylla* and *H. serrata* flowers were often painfully folded and curled, with some damaged beyond recovery. Irrigation made no difference as the cause was petal surface temperature. By contrast, *H. arborescens* cultivars such as 'Annabelle' and its pink derivatives were relatively unperturbed, if not defiantly unbowed and perky.

Though there had been various 'sports' – genetic mutations – found earlier in the wild, as well as seedling variations that had admirable horticultural merit, the plant now called 'Annabelle' was first found in 1910, then lost, then rediscovered in the 1990s in a private US garden by Joe McDaniel of the University of Illinois.

Like all the species/cultivars that flower on the current year's growth, 'Annabelle' can be pruned annually right to the ground like an herbaceous plant, in response to which hard annual pruning it will, in good conditions, produce great balls of cream/white sterile flowers of exceptional size, up to a foot across. It is exceptionally vigorous, suckering and layering itself as it goes. I planted a single plant from a 3ltr pot twenty years ago, which with full light overhead but high shade from an oak from the hottest midday sun, no root competition and, therefore, a good moist root run, had spread to cover an area of over 16m^2 (172ft^2), eventually requiring a mini-digger to prevent it from terminally mugging its companion plantings.

Annabelle's enormous flowers unfortunately will often bend and flop on stems that cannot support their weight, especially when wet. Some describe them as coarse, vulgar and gross, while others take the opposite view; as Michael Dirr puts it in his 2004 *Hydrangeas for American Gardens*: '… garden cognoscenti say the [Annabelle] flowers are too large, gauche, obtrusive; one person's favourite is another's bane; life is great'. I have found that if left unpruned and simply deadheaded, corymb sizes are reduced, the smaller flowerheads better supported, and with a flower on every shoot *en masse* the flowers more or less hold each other up to create an arresting overall spectacle throughout the summer, from June well into the autumn. Whether pruned for large or small heads, the individual florets that make up the corymb typically remain quite small, only 1–1.5cm (½in) across, are greenish yellow as they open, mature to creamy white, then tint green again with age. There is another cultivar on the market that is the spitting image of 'Annabelle', called 'Bounty'. I grow them both, and can find no difference between them, certainly in the garden landscape – a view shared by others.

The scramble for a pink 'Annabelle'

If 'Annabelle' was a breakthrough cultivar that demonstrated the substantial horticultural potential of variations in *H. arborescens*, it set in train a competitive scramble to produce a pink 'Annabelle'. There were one or two forms in cultivation with pink fertile flowers that in themselves had no great decorative quality, but perhaps carried the genes to potentially deliver the ten-million-dollar plant sought by growers and breeders.

Already available were **'Eco Pink Puff'**, **'Pink Pincushion'** and **'Wesser Falls'**: all had some form of pink, in either bud or flower; but all were without significant ray flowers, instead offering packed corymbs of small fertile flowers: overall not sufficiently showy for real popular appeal.

'Pink Pincushion', pink but not ornamental enough to be a popular choice.

'Invincibelle Spirit', thriving in the Sir Harold Hillier Arboretum in Hampshire.

The first new hybrid to hit the market with pink sterile mophead flowerheads in the 'Annabelle' vein was called **'Invincibelle Spirit'**. The buds are a quite deep-pink, opening to a genuine light-pink, then fading with exposure, particularly in full sun. Some shade will help to preserve the flower colour. It is not as vigorous as 'Annabelle', nor are the flower heads as large, but the relatively slender stems can unfortunately still arch and flop. I have found it to have a weak constitution with rather spindly shoots and some dieback. Having said that, irrigated plants in the Hillier Arboretum are growing well in some shade with shelter from the midday sun and leafy topcover, and look happily attractive. 'Invincibelle Spirit' grows to about 1m (3ft) high and roughly the same across. It has some continuity of flower throughout the summer and into autumn, so there can be a nice mix of faded pale and deeper pink fresh flowers at any one time. 'Invincibelle 2' is said to be an improvement in colour and poise, but I have not yet seen it in gardens in England, unless it has been renamed in the somewhat confusing ragbag of pink this and that.

Once other cultivars began to emerge that appeared to be more effective garden plants, a whole new peal of Annabelles appeared, ringing the changes on a dominating pink theme. From **'Sweet Annabelle'** through **'Magical Pinkerbelle'** to **'Candybelle Bubblegum'**, the names say perhaps more about the perceived market than the plants themselves, but competition for the *numero uno* pink version has produced some excellent results. The nomenclature crowds in and can be quite complicated, and rather confusing, involving a prefix and suffix mixture of Invincibelle, Incrediball, Annabelle, Magical, Candy and so on. I have grown four of these later selections for only three years, so have no direct experience of their longer term performance, but can say so far that I have been impressed with them as garden plants, especially through recent record-breaking European summer temperatures. They appear to be as heat-resistant as cold-tolerant, and while some shade may be desirable to preserve the flower colour, they will perform happily in full sun. Colourwise, they are almost all in the same mould – a bright pink in bud and on first opening, then fading gradually to pale pink and thence to greenish/beige. The flowers are in typical 'Annabelle'-style: congested corymbs of sterile mophead florets that tend to flop with their weight, especially in the rain.

My preferred variety to date for both colour and form is **'Invincibelle Ruby'**. It is a compact plant up to about a metre, bushy, neat and rounded in habit, perfect for a small garden. The flowerheads are smaller than most, only 12–15cm (5–6in) across, but the individual florets are larger at 2.5–3cm (1in) in diameter, and a uniform bright ruby red, more red than pink, with the first flowers opening at the end of June here in the UK. There is some flower continuity, so faded greenish flowers are interspersed with bright fresh magenta blooms, which keep the display lively during late summer.

Hydrangea arborescens 'Invincibelle Ruby': the most compact and darkest of the new pink versions.

Hydrangea arborescens 'Magical Pinkerbelle', with 'Annabelle' in the background: good continuity, with fresh flowers into the autumn.

Hydrangea arborescens 'Sweet Annabelle', with large, impressive pink corymbs.

Hydrangea arborescens 'Pink Annabelle', showing a faded and a fresh flower.

The two 'Candybelles' – Marshmallow (left) and Bubblegum (right) – showing the different colour tones.

'Sweet Annabelle' has flowers to rival its white namesake in size, a good 25–30cm (10–12in) across, with individual florets averaging 2cm (1in). Buds are a rich reddish pink, opening deep pink and fading at maturity to a greenish/beige. It makes a strong shrub to 1–1.5m (3–5ft), flowering from June well into August, even in extreme heat.

'Pink Annabelle' appears to be similar to 'Invincibelle Spirit' and may be a synonym, but with rather smaller flowers than the sweet version. All these plants are PBR protected, with 'Pink Annabelle' designated Naha1 and 'Sweet Annabelle' Naha 4.

'Magical Pinkerbelle', an upright plant about 1–1.2m (3ft) high, has corymbs about 20cm (8in) across, with individual florets at 1.5cm (½in) in diameter, so all round a little smaller. This does not detract from garden impact and it seems to go on producing later fresh flowers with greater freedom than most.

There are two Candybelles, **'Candybelle Bubblegum'** and **'Candybelle Marshmallow'**, certainly names to conjure with. Both are plants of about 1.2m by 1m (3ft), with a neat compact habit. The former is described as bubblegum pink, so a medium light-pink, if bubblegum is still same the colour as when I last used it 75 years ago; the marshmallow version is white with a watercolour wash of palest salmon, fading to ivory. Both looked attractive additions when grown in ideal conditions in my local nursery, Signature Hydrangeas in Kent.

Other white mophead forms of *Hydrangea arborescens*

Apart from 'Annabelle', there are a number of other sterile mophead forms of white-flowered *H. arborescens* worth consideration, each different in style and presentation.

One is a recent introduction called **'Incrediball'**, developed specifically to have strong stems with sufficient resilience to hold the heavy, football size, 'Annabelle'-style flowers erect. The punning name well describes the plant's major characteristic and is synonymous with 'Strong Annabelle'.

The unfloppable 'Incrediball'/'Strong Annabelle' has stout stems that keep its head up.

Hydrangea arborescens 'Grandiflora': an old variety but a good companion for summer roses.

'Grandiflora', an old variety introduced as long ago as the late-1800s, is quite different in style and character to 'Annabelle'. 'Grandiflora' has an irregular, flatter, lumpier inflorescence composed of bigger individual florets with a more informal look. Like 'Annabelle', the stems are weak and, typical of most other *H. arborescens*, it has has a floppy habit, with suckers and layers spreading to 2–3m (6–9ft) over some years; but with the right companions in support it can create a gently undulating low wave of foamy white suffused with pale green, just under a metre tall. I grow it with complementary – and complimentary – brightly coloured shrub roses, and it provides a background for their pinks, purples and reds, while nonchalantly leaning on and poking through them, in a bouquet effect. Both the pale-green foliage and the all-summer white-green flowers are a perfect foil for dramatic spots of colour. But do not plant it if you are a strict disciplinarian.

'Sterilis', a form of ssp. *discolor*, and formerly known as *H. cinerea* 'Sterilis', is different again, with smaller, more regular and more symmetrical ball-shaped corymbs, each composed of smaller individual florets. It too has a straggly habit, and the stems can look a bit bare and bony with age. There are large plants in some old gardens in the UK, but it is no longer much planted. It lacks the personality of 'Grandiflora' and is less effective in the garden.

Two recently introduced dwarf forms in 'Annabelle' flower-style of which I have no personal experience are 'Invincibelle Wee White', which makes a continuously flowering small mound of up to 0.5m (1½ft), and 'Invincibelle Limetta', which is a little taller at about 1m (3ft). The flowers open lime green, pass to white, then mature to a clear jade green.

New white lacecap forms

Another recent introduction is *H. arborescens* **'Hayes Starburst'**, which appeared as a chance seedling in the garden of one Hayes Jackson, in Illinois. It also has flowers that typically

Hydrangea arborescens 'Hayes Starburst'.

start green, transition to creamy white, then age greenish again, giving a loose, cool, watercolour wash effect. But in this case, the individual florets making up the rather loose corymbs are double and have narrow small, spiky, star-like petals.

'Hayes Starburst' grows to around a metre, and although it flops about, it also sends up strong individual shoots capped with these strikingly-petalled flowers. For me it works well as a ground cover, spreading loosely in the relative shade and shelter of companions like *Deutzia* and *Viburnum* to provide colour and eye-catching texture over a long summer. I found it unco-operatively slow to start, but reliable once settled down.

I believe **'Mary Nell'** is not yet available in Europe, but it looks like a very beautiful plant, halfway between a mophead and a lacecap, with a peripheral ring of multiple overlapping ray flowers surrounding a central zone of cream fertile flowers. It is another introduction by Joe McDaniel, named for his wife, in itself a recommendation. This also offers a low-growing palette of whites, creams and greens as ground cover or as background for seasonal interest and contrast with more brightly coloured companion plants.

These *H. arborescens* novelties are coming on to the market through garden centres thick and fast at the time of writing, all of an uber hardiness not generally associated with hydrangeas in the public mind. Producers claim hardiness temperatures down to –30°C. It would be well worth keeping an eye open for them, especially if you are in a cold spot, say in Scotland, where some of the more fastidious or winter-susceptible forms of *H. macrophylla* might not perform satisfactorily or even be viable.

Other names to conjure with and already listed in the USA but which I have not seen in Europe to date (but will no doubt appear soon if thought to have merit) are 'Bella Anna', 'Invincibelle Garnetta', 'Incrediball Blush', 'Terry Greer', 'Green Dragon' – the list goes on, and by the time you are reading this book, there are likely to be more, as forms and hybrids of this American native species continue to gain popularity as easy plants: a relaxed, everyman garden plant to rival *H. paniculata* for hardiness and flowering capacity. These new *H. arborescens* cultivars are ideal for cold gardens looking for reliable summer colour.

Hydrangea quercifolia

This is quite unlike its sister species in the Americanae subsection: as the name implies, it has leaves like an oak, notably bigger, but

Hydrangea quercifolia is the State flower of Alabama.

rather like the big bold leaves of the local *Quercus rubra*, and also bears its flowers in terminal panicles like the Japanese *H. paniculata*. (These are the only two species in the whole genus to sport large panicles of flowers at the end of shoots, perhaps another testimony to the link between the American and Asian floras.) It does, however, have in common with *H.* arborescens an eastern/southern US distribution, and in places it grows together with *H. arborescens* in the woodland understorey.

A unique endemic species

Hydrangea quercifolia is celebrated as the State flower of Alabama; its range extends further to the Carolinas, Georgia, Florida, Louisiana, Missouri, Mississippi and west to Tennessee. Typically, it will reach some 2m (6ft) by perhaps 3m (9ft) across, though there are now much smaller dwarf forms in cultivation. The young shoots and petioles are densely covered with reddish hairs; the leaves are lobed like large oakleaves, some up to 20cm (8in) long, five-lobed with the lobes coarsely toothed and the underside hairy. The fertile flowers are white, packed into a pyramidal panicle, usually 15–25cm (6–10in) long and wreathed, sometimes densely, by conspicuous white ray flowers, some 2–3cm (1in) across, some ageing to pink. They appear from July to October. The autumn colour, though it can be patchy in the UK's capricicus climate, is a major feature of the species, with dark wine purple, sometimes mixed with red, staining the bold look-at-me foliage.

Leaf shape and autumn colour are two key features.

Performance linked to climate

The *H. quercifolia* distribution in the southern states readily suggests that it enjoys significant warmth during the growing season and this perhaps explains why it has in the past been regarded with some suspicion as not reliably hardy in dank and cold UK and Europe. Haworth-Booth writes: 'It is a pity that this handsome hydrangea is just too tender to thrive in any but favoured gardens in the south and west of England.... It just fails to be really decorative... unless the season is an unusually favourable one...' by which I infer he means when the summer is very warm and prolonged, the sun shines a lot and the livin' is easy.

I have certainly found that *Quercifolia* cultivars will flower much better in full sun and have not found them much damaged by winter frost when such sun can ripen and harden the wood. But then we have had a recent prolonged run of mild winters, so current performance may not make a fair comparison with the Howarth-Booth conditions some 70 years ago.

My own experience has not been over-enthusiastic about the garden performance of some half-dozen cultivars, and with one or two exceptions, there has been a general reluctance to flower sufficiently freely to make a big impact. They do not appear to be good shade bearers in this climate, growing freely but flowering parsimoniously, with some flowers hidden under copious foliage. Some have had winter bud damage and with a little dieback eventually lackadaisically dwindled away to the point where they no longer paid their rent.

Cultivars are very popular in the US and widely planted; but in the UK, certainly in the past, *H. quercifolia* has been rare in gardens and only a handful of cultivars have been on offer in the trade. There is some evidence that this may be changing; over the last ten years some twenty cultivars are listed in the RHS Plant Finder as available from UK nurseries, with cultivars like *H. quercifolia* 'Alice', 'Snowflake' AGM, 'Snow Queen' AGM, 'Harmony' and 'Pee Wee' maintaining their relative popularity with nurseries, a useful index of public approval; the species itself is listed as widely available (stocked by more than twenty nurseries). To this should be added numerous garden centres. It may be that the changing climate is having a beneficial influence on the garden effectiveness of *H. quercifolia* in unpredictable but warming summers and more gardeners are now planting it.

Some reliable *Hydrangea quercifolia* cultivars

Hydrangea quercifolia **'Alice'** has grown in my garden to some 2.5m (6ft) tall, making a sprawling mound about 3m (9ft) across, in full sun. The flowers are produced more or less

Hydrangea quercifolia 'Alice'.

Hydrangea quercifolia 'Harmony'.

continuously from June onwards, but never covering or colouring the whole bush. They turn a nice pink as they mature, then wrapping-paper brown, and I remove these during flowering as they spoil the overall effect. Panicles are large, up to 30cm (1ft) long, studded freely with showy ray flowers.

Hydrangea quercifolia **'Harmony'** is a unique form, found in a churchyard in Alabama some 40 years ago. It is not an easy plant, the rather weak stems finding it impossible to hold up the heavy flower heads. These are a heavy agglomeration of overlapping, quite small, greenish-white florets, packed into a large, rather uneven and lumpy, broad mophead panicle. My young plant presented as a ring of panicles resting on the ground around the plant, flopped out from the centre. It is worth taking trouble with it as it is a striking plant when well grown; a robust stake to support its weak stems in the early stages can be helpful. It can make a mound up to 1.5m (5ft) treated in this way.

Hydrangea quercifolia **'Snowflake'** is a selection with 'double' flowers, with florets forming one on top of the other to give a spiral effect, like small pieces of paper on a spike, very similar to double forms of *H. involucrata*, a feature the Japanese call '*yae*' (*see* Chapters 5 and 6). It grows to about 1.5m (5ft), a spreading shrub, exceptionally freeflowering with heavy, tumbling panicles of sterile white florets that turn reddish/pink on maturity. Autumn colour is a mix of red/purple, usually lightened by orange and red tones. It is a first-class flowering shrub of robust constitution that enjoys plenty of sun.

Hydrangea quercifolia **'Snow Queen'** is not so robust and I have lost my plant, which declined over time, perhaps through less than ideal drainage. However, when happy, it has stronger stems that hold its flowers up nicely, some fully erect. It makes a rather open plant up to 2m (6ft) with panicles up to 20cm (8in) crowded with large, plentiful sterile ray flowers.

Hydrangea quercifolia 'Snowflake', early (left) and late (right).

Hydrangea quercifolia 'Snow Queen'.

Hydrangea quercifolia 'Sykes Dwarf'.

There are a few cultivars suitable for a small garden. *Hydrangea quercifolia* **'Pee Wee'** is perhaps the best known and is a compact 1 × 1m (3 × 3ft), with a neat broad pyramid of mainly sterile, not too congested, ray flowers, about 10cm (4in) long and 8cm (3in) across the base. Autumn colour is the typical wine red. I grew *H. quercifolia* **'Sykes Dwarf'** for some years but it didn't thrive once overshaded by growth put on by nearby trees. It too is about 1 × 1m (3 × 3ft), with flowers similar to 'Pee Wee', but if anything, slightly larger.

Hydrangea quercifolia **'Munchkin'** is a more recent introduction, a little bigger at about about 1.2m (3ft), a relatively compact domed bush with a neat open panicle of mixed fertile and white ray flowers, turning pink as they mature.

Hydrangea quercifolia **'Ruby Slippers'** is similar in height and habit, but the flowers change from white to pink, then gradually turn a deep rich red.

Hydrangea quercifolia **'Tennessee Clone'** is another I have lost in cultivation, a 2m (6ft) plant with large broad panicles over 20cm (8in) long, from seed collected in Tennessee by the de Belders in the1970s. The white flowers fade to pale green.

There are several other forms listed and no doubt new forms will continue to appear; they seem to come and go, but *H. quercifolia* 'Snowflake', 'Harmony', 'Alice', 'Peewee' and 'Ruby Slippers' are tried and trusted and have stayed the course.

Hydrangea quercifolia 'Munchkin'.

Hydrangea quercifolia 'Tennessee Clone'.

CHAPTER 4

HETEROMALLAE: THE ASIAN STALWARTS

There are just two species in this subsection: *H. heteromalla* from the Himalaya and China and the increasingly popular *H. paniculata* from Japan and east China.

A plant for anywhere

On a ridge of the Nushan on the Mekong/Salween Divide in W. Yunnan, China, at around 3,500m (11,500ft) in 1994, I spotted two small trees, both hydrangeas, growing in stark exposure and overlooking a long, damp valley covered as far as the eye could see in the massed pink flowers of dwarf *Rhododendron calostrotum*, threaded with streams and rivulets and framed by a backdrop of weeping junipers on the drier slopes. These were both specimens of *H. heteromalla*, a species found at a higher altitude than any other hydrangea in China or, indeed, elsewhere and, consequently, exceptionally tough and durable.

On the same mountain, but much lower down, was *H. aspera*, but as altitude increased, *H. aspera* eventually disappeared before the first *H. heteromalla* was spotted. I have never seen any introgression in the wild between the two, via any hybrid seedlings – but rather have always noted this altitude gap between them. They may be incompatible: I have not tried hybridising them. If it did turn out that they were compatible, the cross might just produce an exceptionally tough, hardy hybrid for universal planting with good, pink flowers.

Their natural occurrence at relatively high altitude in the cold temperate zone means that, along with the related Japanese

An old veteran *H. heteromalla* on the Nushan, at 3,000m (10,000ft) on the Mekong/Salween divide, Yunnan.

OPPOSITE PAGE: An outstanding form of *H. heteromalla* from the Himalaya found by plantsman David Clulow, the poise of its pure-white flowers allied to its reliable hardiness making it an excellent garden plant.
H. paniculata 'Vanille Fraise' on the edge of woodland at RHS Wisley and demonstrating the gradual colour change from white to red of many cultivars, to create a bicolour effect.

Hydrangea heteromalla 'Bougies' survives sub-zero winters in Finland.

H. paniculata, *H. heteromalla* is among the hardiest of all Hydrangea species. Indeed, *H. paniculata* and *H. heteromalla* are so closely related that a recent revision (2016) made by two Japanese botanists, Ohwi and Hakiyama, based on laboratory molecular phylogenetic analysis, merged them as a single species.

Any plant from the Heteromallae subsection will be a hydrangea for anywhere, able to flourish in the very coldest garden. Indeed, a form of it grows in Finland, where the mean annual minimum winter temperatures away from the coast usually fall to around −28°C and, occasionally, below −30°C. The Mustila Arboretum near Elimaki in Southern Finland, for example, grows its own selection called *H. heteromalla* 'Bougies', with double flowers. It will take full frost exposure in its stride, and flowers most freely in full sun, which also produces rather smaller leaves, often tinted bronze or red.

Hydrangea heteromalla distribution and classification

In nature, *H. heteromalla* ranges widely from Nepal along the Himalaya into west and north China. It is frequent along the high divides separating the three great rivers of that western region: the Salween, the Mekong and the Yangtse, falling south off the Tibetan plateau.

Those of Himalayan provenance are said to be less hardy than the Chinese plants, but to date I have not found this to be the case.

Like most hydrangeas, *H. heteromalla* is highly variable in habit, foliage and flower, and various forms have been selected and given full species or varietal status by botanists over the

Hydrangea heteromalla in the wild: Nepal, a Crug Farm collection.

Hydrangea heteromalla in Sichuan in full flower (above), and in autumn colour (below).

years. To illustrate this variability, no less than twenty entities formerly regarded as separate species, varieties or forms were all sunk into *H. heteromalla* as synonyms by McClintock in her 1957 monograph. Before that, Howarth-Booth had listed thirteen separate species in the Heteromallae section in his 1950 review, and although many of these were not then in cultivation, all have now disappeared from most current lists. Only a few of these 'species' were distinctive in flower, and many not sufficiently decorative to be first choice for a prominent place in gardens. Nonetheless, there were discernible differences in leaf shape, vestiture, size, bark and habit, which, added to the general propensity on the part of early explorer botanists to add something excitingly 'new' to their herbarium press, led to this initial proliferation of 'species'.

Hydrangea heteromalla in the wild: Bhutan.

Separate species listed by Haworth-Booth (1950) now sunk into *Hydrangea heteromalla*

- *H. xanthoneura*
- *H. dumicola*
- *H. mandarinorum*
- *H. bretscheideri*
- *H. hypoglauca*
- *H. macrocarpa*
- *H. pubinervis*
- *H. tsangii*
- *H. stenophylla*
- *H. lobbii*
- *H. pubiramea*

McClintock sank these and altogether some twenty species and varieties into *H. heteromalla*.

An excellent compact SICH form of *H. heteromalla*, found on the Kew expedition to Sichuan in 2004.

Hydrangea heteromalla forma *dumicola* is typical of the descent into anonymity of these various forms, some of which could still contribute usefully to the garden scene. It was originally introduced as a species by George Forrest from west Yunnan, but is now sunk within *H. heteromalla* as simply a form.

It was distinguished by a rough bark and rather large, pale-green leaves with a whitish underside composed of strigose white hairs. It was growing in Maurice Mason's garden at Larchwood in Norfolk as a small tree. Its genuine claim to distinction is that it had definite built-in ambitions to be a proper single-stem tree, and would doubtless make an attractive early flowering lawn specimen if encouraged to form a single trunk, with a gently spreading semi-weeping crown studded with flowers on a small tree perhaps a bit over 4m (12ft) tall. Sited like this, it would be effective over a long period, and an attractive and unusual feature. It is now rarely seen in gardens. It was valued by the late Princess Sturza in her garden at Vasterival near Dieppe.

The type species

The typical inflorescence of all forms of *H. heteromalla* is white, with both the four-sepalled ray flowers and numerous fertile flowers prominent in an open, branching corymb.

A keen nose can detect a sweet fragrance. But from white, with time and light, the ray flowers turn over and the action of light on the reverse turns them into a uniform pink or even plush red, creating a relaxed, brushed mix sometimes suffused with greens, creams and pinks, an effect lasting well into autumn.

With some more recent introductions having bigger and showier flowers, it is worth keeping an eye open in gardens and nurseries for these improved forms, which may not be named,

Two examples of *H. heteromalla* in the wild in Yunnan, China.

An old *H. heteromalla* at Westonbirt Arboretum, around 9m (30ft) tall.

but are well worth propagating. Bleddyn and Sue Wynn Jones of Crug Farm, for example, the most prolific British plant collectors in East Asia of our time, list six collections from the wild in their catalogue.

A notable advantage of the species is that it generally flowers early in the hydrangea season, covering itself with prominent white lacecaps as early as May. Some forms are the first hydrangeas to flower in any collection, along with the smaller and unrelated *H. luteovenosa*. With little competition from other hydrangeas at that time, it is conspicuous in the late-spring garden, the pure white corymbs contrasting sharply with its dark foliage.

Hydrangea heteromalla foliage varies in size, with some forms having attractive leaf characteristics – for example, ovate leaves, acuminate with a long, elegant point, ranging from 7 to 15cm (3–6in) long, dark green above but greyish below, often tinted purple or red and sometimes with prominent red petioles. Leaf size variation is also highly dependent on exposure and shade: tending to be smaller with more sun, and larger with more shade. These leaves can often clothe stout, stiff, brown stems creating anything from a small 1.5m (5ft) bush to a monster 9m (30ft) multi-stemmed tree, such as that long in cultivation at Westonbirt.

Some garden cultivars

There are now a number of quite compact clonal selections in cultivation regarded as eminently gardenworthy. The Esveld nursery in the Netherlands, for example, known *inter alia* for its large collection of hydrangeas, lists some fourteen 'cultivars and subspecies' of *H. heteromalla* that they clearly deem marketable, and thus meriting a place in gardens.

Hydrangea heteromalla var. *xanthoneura* at White House Farm.

Hydrangea heteromalla 'Jermyns Lace'.

Their list interestingly does not include what is now designated in the RHS Plant Finder as the **Bretschneideri Group**, a 'group' moniker usually taken to suggest a degree of acceptable variability from seed, but with over-ridingly similar characteristics. The main distinction of this group from other forms is its bark, which exfoliates into chestnut-brown flakes and strips, while other Heteromallae have tight bark, sometimes lenticillate like birch, and varying from grey to red-brown. *Hydrangea heteromalla* 'Bretschneideri' is an old, well-tried variety, based on seed sent to Germany by Dr Bretschneider from Beijing in the 1880s. Fully hardy, in the open it is a medium-sized shrub growing up to 2.4m (8ft), with 15cm (6in) corymbs studded with small ray flowers in the second half of June.

Closely allied is a variable plant formerly known as *H. heteromalla* **var. *xanthoneura***, with rather larger corymbs and pure bright-white ray flowers some 5cm (2in) across. But this is a much bigger plant, reaching over 9m (30ft) in cultivation in favourable conditions.

Hydrangea heteromalla forma *wilsonii*; showing the scale and elegance of corymb.

Another form of *H. heteromalla* var. *xanthoneura* in which the white ray flowers turn pink with age, selected and grown at the Hillier Arboretum, is **'Jermyns Lace'**. This is a vigorous shrub to about 4m (12ft) with larger white lacecap flowers borne in June.

'Nepal Beauty' was introduced by the Esvelt nursery in Boskoop, Holland in 1994. It is distinguished by its larger, red-tinted and red-margined foliage and larger flowerheads appearing rather later, in July. It is listed as a hybrid but without information as to parentage, presumably unknown by the nursery.

Collected in the Himalayas and grown for years at the Hillier nursery as *H. robusta*, a subspecies of *H. aspera*, *H. heteromalla* **'Snowcap'** is one of the finest heteromalla cultivars, with large leaves, an upright habit and large, white corymbs up to 20cm (8in) across. It is tolerant of exposure and indeed will flower extra freely in full sun.

Hydrangea heteromalla **forma *wilsonii*** is now very rare in gardens but remains one of the best forms. It is one of the first hydrangeas to flower, often opening in early May. It is easily pruned into a small tree up to 3m (9ft).

Hydrangea heteromalla **'Morrey's Form'** is rather similar, but bushier up to around 2.6m (12ft) with white, narrower and curled ray flowers. And with seed collected on the highest mountain in N. Vietnam by the Wynn-Joneses, *H. heteromalla* 'Fan Si Pan' is distinguished by its bronze young foliage, borne on red shoots with prominent red petioles. In the wild this cultivar makes a medium-sized shrub up to 3m (9ft) in height.

The Wynn-Joneses of Crug Farm Plants list five further collections made in India, Nepal, Vietnam and China. The only way to decide if any of these would be your own planting preference is to visit Crug Farm or other demonstration gardens and see them in growth and flowering.

Hydrangea paniculata: the universal crowd pleaser

After *H. macrophylla*, the cultivars of *H. paniculata* are currently the most popular of all hydrangea species in terms of both sales and cultivars available. Paniculatas are, in fact, more widely planted now geographically than *H. macrophylla*, thanks to their rugged constitution and relative indifference to soil types and testing winters.

I was invited to Nova Scotia in 2018 to give a talk about hydrangeas in cold climates. Prior to this I had assumed that hydrangeas and Nova Scotia were something of a contradiction in terms, because winter temperatures in eastern Canada persist at daunting subzero levels for many months, but my hosts insisted a lecture on the genus would be of interest. As it turned out, I learned a lot about hydrangeas from them. What I had forgotten to take into account was precisely the tough qualities of *H. paniculata*. One of my abiding memories was driving through residential neighbourhoods seeing specimens of *H. paniculata* 'Grandiflora' everywhere, from front gardens, to allotments, to urban green civic spaces. Instead of being pruned down hard to produce the great overblown, congested, sagging, football-sized flower-heads commonly seen in Europe, these plants were invariably treated in an opposite fashion, pruned up into small trees. Some were clearly of great age, wearing that aura of relaxed dignity that only age confers. *Hydrangea paniculata* 'Grandiflora' was introduced from Japan by von Siebold in 1870, and appears to have flourished in north-eastern Canada ever since, typically standing on a single stem with exfoliating bark and a graceful crown, gently weighed down with many relatively small flower-heads, in an informal, semi-weeping picture of some elegance – and this in average winter temperatures well into the minus twenties.

Hydrangea paniculata in nature

Hydrangea paniculata is essentially a Japanese species, found in Kyushu, Honshu and Hokkaido, but spilling over into Sakhalin, eastern and central China and Taiwan. It can make a small tree as tall as 8m (24ft) or quite a small shrub of 1.5–2m (5–6ft). It is bone hardy, reportedly down to –30°C and tolerant of all soils and situations. It will tolerate relatively dry conditions, though in cultivation, like all hydrangeas, it prefers moisture during the growing season to optimise flowering performance. It is happy in exposure, and will tolerate full sun. Indeed, it is unhappy in too much shade, and will react by producing blind shoots.

Hydrangea paniculata is one of only two hydrangea species to flower in terminal panicles, the other being the American *H. quercifolia*. In foliage, flower and leaf and shoot vestiture, it is much less varied than its close relative *H. heteromalla* and is altogether a more uniform plant, though the wild forms can vary in the openness and looseness of the flower panicle and in the number of sterile ray flowers, some with long pedicels, creating a more open, delicate, looser panicle, notably in Taiwanese and Chinese populations.

Hydrangea paniculata 'Grandiflora' is frequently seen in Nova Scotia, where winter temperatures regularly dip below –20°C.

A typical wild form of *H. paniculata*.

An attractive wild form of *H. paniculata* with pretty pink buds.

The open airy panicle of *H. paniculata* in Hunan province, China.

Hydrangea paniculata 'Pinky Winky' showing lower mature flowers bright red, with those at the tip still in white bud.

As some indication of its relative uniformity, McClintock sank only three related separate entities into the type species, each from outside its main distribution area in Japan: two from China and one from Sakhalin.

Left to its own devices and unpruned, in cultivation *H. paniculata* will make a large shrub up to 6m (20ft), upright at first, then with a spreading, opening crown. It is one of the latest of the hydrangea species to leaf out, unfurling ovate or elliptic leaves up to 15 × 7cm (6 × 3in) in June, their size depending on shoot vigour and degree of exposure. Leaves are in opposite pairs and occasionally occur in threes. There is little pubescence, and foliage is unremarkable in colour, usually a mid to dark green and only occasionally with a reddish suggestion in leaf and petiole.

The wild-type inflorescence is composed of an abundance of small, creamy, fertile flowers, sometimes with little pink buds, in a pyramidal panicle; this is dotted and wreathed with larger four-petalled, white ray flowers, typically about 3cm (1in) across, the whole forming a usually irregular cone, anything up to 25cm (10in) long, though occasionally rounded or truncated to be an almost flattened corymbiform.

In cultivation many cultivars have only sterile ray 'mophead' flowers with fertile flowers absent; others have a varying ratio of fertile flowers to ray flowers to give a more open panicle. Buds open from the bottom up to the apex in the panicle and, as the flowers age, many acquire pink or red flushes, some becoming uniformly red, others bearing red flowers at the base, with white at the top, still opening and fresh, producing a bicolour effect.

The explosion of garden varieties of *Hydrangea paniculata*

In the last 30 or so years, there has been an explosion in the number of *H. paniculata* cultivars available. From only three mentioned in the 1950s Howarth-Booth book, and despite the de Belders of Belgium subsequently introducing new seedlings, still only eight cultivars were listed in the RHS Plant Fnder in 1987 – the current edition now lists a hundred. Astonishingly, there are now no less than a runaway 140 different cultivars in the UK national collection in Derby. These have all originated as

The first RHS trial of paniculatas at Wisley in 2008, to be followed by a further trial in 2021.

variations from the single white species, without significant hybrids with other species bringing the possibility of fresh colour and character into play. An excellent Royal Horticultural Society Award of Garden Merit trial of *H. paniculata* in 2008 assessed 68 entries representing 47 cultivars, different varieties from the UK, Holland, Belgium and the USA.

By 2019, a further trial was deemed necessary as in the intervening decade, a further 60 cultivars had been introduced on to the market. The first planting of this trial was made in autumn 2021 in the newly opened RHS garden at Bridgewater in north-west England. The photographs are taken in September of the following year, showing one or two first flowers, late in their season. This trial will be an invaluable indicator as to any improvement over existing tried and tested cultivars, and an updated pointer to cultivar choice for gardeners.

The reasons for the extraordinary growth in cultivar numbers are not hard to find. It is driven by a competitive commercial market eager to invest in the perfect everyman plant: easy to grow on any soil, in any climate, in any situation, with specimens to suit any taste, right across the world. The market is global and thus potentially huge. Many thousands of seedlings have been raised by enterprising nurserymen in Europe, but also in the USA and Japan. *Hydrangea paniculata* combines this indestructible character with a long season of effective flowers, many of which transition from white to deep pink or red with time and represent one of the finest summer flowering shrubs in the whole garden pageant, often remaining effective for many weeks.

This combination of benefits, however, has a downside. Because cultivar development is market-driven, there has been a surfeit of new introductions to the point of confusion, with differences so slight they can now flummox even the most discriminating and picky of gardeners. The most obvious differences hinge on main characteristics such as ultimate plant size (smaller gardens demand smaller plants); the style of the inflorescence (open with mixed fertile and ray flowers as in

Hydrangea paniculata 'Big Ben' (top), 'Limelight' (middle) and 'Vanillle Fraise' (bottom) at their best on Battleston Hill at Wisley, with rhododendrons and other woodland shrubs and trees.

nature, or mophead with no small fertile flowers); and while all flowers are essentially white, some turn pink or red with age.

Within these obviously well-defined categories, even a trained eye, accustomed to spotting nice differences, will find it difficult to distinguish some cultivars from others,

Examples of the 'new' paniculata cultivars in the RHS Trial at Bridgewater photographed in 2022. Clockwise from top left: 'Little Fraise', 'Candlelight', 'Fraise Melba', 'Magical Vesuvio', 'Living Colourful Cocktail', 'Early Harry'.

which, especially once transferred from nursery pot to the garden, perform to similar effect. It is easy to demonstrate such uncertainty in cultivar identification by simply removing the label and asking the question. Some of the characters may differ in their response to different circumstances, as they are variously influenced by shade, sun, weather, vigour or pruning. This can only be learned by growing them and living with them.

So which *H. paniculata* to choose? Indecision is the product of excessive choice. Too much too similar choice is as bad as too little. It's where the market risks selling itself in and then selling itself out by overdoing it: one is 'spoiled for choice'. Even if you are happy to make a real effort to filter the best, or simply accept what is available in the nearest garden centre, new introductions are becoming available so thick and fast you will almost certainly be behind the game in any event. And the results of the RHS Bridgewater AGM Trial are not due to be published until 2025.

There is some help at hand. There are garden visits, now, to see for yourself what corresponds best with your own tastes and preferences. The established and well-managed collection on Battleston Hill at RHS Wisley, in Surrey, for example, while not telling the whole story, would be an excellent start, particularly in demonstrating the use of paniculatas in a mixed, mainly woody, planting; plus of course the national collection of paniculatas in Derby. On the continent, plants can be viewed at the Esveld nursery in Boskoop in Holland, at the Shamrock collection at Varengeville in Normandy in France, and the Belgium National collection in Destelbergen.

Dipping into the options and making my own personal suggestions of some of the most gardenworthy sorts is not satisfactory, because with such a wide choice it will always come down to individual taste and inclination. You can enjoy these long-lived plants for a lifetime, so it's good to get your cultivar choices right *ab initio* if you can.

The utility of RHS Award of Garden Merit (AGM) trials

As a reliable guide, I would also enlist the help of the results of the excellent RHS Award of Garden Merit trial of paniculatas in 2008. The full report is online. This is only one of several *H. paniculata* trial results available in both the UK and Europe. The 2008 RHS trial was exhaustively assessed by a phalanx of knowledgeable but disinterested and objective cognoscenti from the RHS Woody Plant Committee, over no less than four years, with four or five three-hour visits per year, each followed by discussions sifting the evidence and arguing about which cultivars should be selected as achieving a level of excellence as garden plants to qualify for an AGM award. As always, some were self-selecting as 'givens', others were subject to lively discussion, while those on the margins required a brutal chairman.

The objectives of the RHS 2008 *Hydrangea paniculata* trial

Trial objectives:

- To compare and assess new and old cultivars.
- To recommend the Award of Garden Merit (AGM) to those considered to be the best.
- To determine correct nomenclature.
- To obtain specimens, photographs and descriptions as a record to be held in the RHS Herbarium at Wisley.

Judging criteria:

- Habit
- Bud, flower, faded flower
- Floppiness of head
- Foliage

To be recorded over time:

- Habit
- Response to pruning
- Stem colour
- Foliage colour
- Flowering dates
- Shape of inflorescence
- Early, mid-season and late-season colour of flower

Pruning – an important aspect of the trial

Three specimens of each plant to be trialled were planted, and pruned in the following year. The front plant of each entry was cut back hard to two buds, the middle plant lightly pruned to four buds and the back plant simply deadheaded. This pruning regime was one of the most interesting aspects of the trial. It became evident that pruning could be used to manipulate the flowering time of many cultivars, as well as the size of the flower head.

A simple minimum deadheading as early as January could produce an early flowering, with more numerous and smaller flowers. At the other end of the scale, pruning hard at about bud break at the end of March or in early April would postpone flowering by over a month, as well as producing massive flower heads. A moderate, medium prune in late winter unsurprisingly produced a result somewhere in between, and often proved to be the best solution for general garden decoration.

If a *H. paniculata* plant gets out of hand, you can simply take it down and start again. I took a winter chainsaw to a 23-year-old plant of *H. paniculata* 'Phantom', reducing it to less than 1m (3ft), and by August it had produced enormous heads some 0.3m (1ft) long and the same across, on stiff,

The Wisley trial, showing pruning methods: (front row) hard pruned, (middle) medium and tall (back row) simply deadheaded. All four images on this page appear courtesy of the RHS.

Hard pruned (left); medium pruned (middle); deadheaded only (right).

strong new shoots. Once established, paniculatas are hard to dethrone. They can easily put up with mismanagement or neglect.

Some *Hydrangea paniculata* cultivars with the AGM award

In the trial, five plants gained the Award of Garden Merit, plus two more, subject to availability (a condition of the award is that they are currently on the market). In addition, two cultivars retained an existing AGM award, giving a total of nine winners from the total entry of 47 cultivars. It was quite an exacting test.

One of the outstanding plants in the trial was *H. paniculata* **'Limelight'** AGM. This has stunning lime-green flowers on opening, maturing to cream to white, eventually becoming a mix of green and pink as flowers fade. Light shade intensifies colour. The panicles are closely packed with only sterile ray flowers, roundish and short. Light pruning is advised to get the best effect: it limits the size of the heads, with large heads from hard

Hydrangea paniculata 'Limelight'.

Hydrangea paniculata 'Big Ben'.

pruning having a tendency to hang or flop. Flowers are freely borne on a vigorous bushy plant.

Hydrangea paniculata **'Big Ben'** AGM certainly has a panicle the shape of the Westminster tower, with the apparent ambition to be the same size. It is very free-flowering on strong, red stems. The tall panicles have a nice open, well-balanced mixture of greenish-white fertile and ray flowers with some fragrance. The latter mature to a deep pink, turning over on long pedicels. It is one of the best of the 'natural' non-globose varieties that is akin to the wild species.

Hydrangea paniculata **'Kyushu'** AGM had its award confirmed following the trial. Its main advantage is its neat and moderate habit, to about 1.75m (5½ft), making it useful for a small garden. It also flowers early, especially if unpruned and only deadheaded. It was introduced by Collingwood Ingram from Kyushu in Japan, and distributed by the de Belders. It is a good form of the wild species, with the relatively scattered ray flowers remaining white amid a mass of cream fertile flowers on a quite small panicle. For me, the award was borderline.

Hydrangea paniculata **'Pink Diamond'** had its 1993 AGM reconfirmed after the 2008 trial. This makes a big bush potentially over 2m (6ft) tall and wide, with long, cone-shaped panicles bearing white ray flowers that later change to a vibrant pink, providing wonderful late colour. This is a de Belder introduction and one of their best.

Another cultivar suitable for small gardens is *H. paniculata* **'Pinky Winky'** AGM, perhaps the most dramatic of the bicoloured varieties, with the base florets turning a deep reddish-pink progressively as the flowers age from the bottom up, and with distinctive, strongly tapered, pointed panicles. Flowers last

Hydrangea paniculata 'Kyushu' at the Annapolis Royal Historic Gardens, Nova Scotia.

Hydrangea paniculata 'Pink Diamond' AGM.

pretty well before browning, on a neat bush, upright, with stiff dark red stems.

Hydrangea paniculata **'Silver Dollar'** AGM is another compact plant, with heavy, dense heads of white flowers, maturing to a pleasant pink. The sturdy stems support the heavy heads well. Also excellent for small gardens.

Hydrangea paniculata **'Phantom'** AGM also has large heavy flower-heads well supported by strong, upright stems, but on a

Hydrangea paniculata 'Pinky Winky'.

Hydrangea paniculata 'Silver Dollar'.

larger bush, freely branching. It is similar to the old *H. paniculata* 'Grandiflora', which it supercedes. It has rounder flower-heads with a nice yellow tip, which do not flop as they are held on stronger stems. An excellent mophead form.

Hydrangea paniculata **'Vanille Fraise'** failed to get an award, largely because prematurely added to the strawberry and vanilla mix was unwanted chocolate, with the red flowers browning at the base before the white top had opened, especially in wet conditions, which rather spoiled an otherwise dramatic and show-stopping effect. When fresh, this plant can have enormous impact, but its freshness does not last long, and

Hydrangea paniculata 'Phantom'.

Hydrangea paniculata 'Vanille Fraise' as a young plant.

Hydrangea paniculata 'Grandiflora'.

the effect is relatively short-lived before the picture is spoiled. It may last longer with some shade. The heavy heads will hang and droop too, so it is advisab e to keep pruning light. I have it pruned as a standard so that it can happily semi-weep.

Hydrangea paniculata **'Grandiflora'**, which had held the AGM since 1993, had it rescinded in 2008, largely because the weaker stems failed to hold up the heavy heads. These can weigh a kilo when wet! It is affectionately known as sheepshead on the continent and pee gee in the USA. It was acknowledged that it improves with age and light pruning.

Hydrangea paniculata **'Floribunda'** also lost a long-held AGM on the grounds that it had been superceded by similar but superior cultivars such as 'Big Ben'. It, too, is close to the type species, but with long, slender, pointed panicles with a scattering of white ray flowers. It is readily pruned up into a nice small tree and would make a nice lawn specimen, flowering as it does of course when lawns are most in use.

Peter Zwijenburg, a celebrated Dutch raiser, as well as achieving AGMs for 'Limelight', 'Phantom', 'Silver Dollar' and 'Big Ben', also had two further cultivars recognised, subject to availability: today both are listed in the RHS Plant Finder. *Hydrangea paniculata* **'Dolly'** is not dissimilar to 'Grandiflora' in habit and size of inflorescence, but carries itself better, with

Hydrangea paniculata 'Floribunda'.

Hydrangea paniculata 'Pink Lady'.

Hydrangea paniculata 'Diamant Rouge'.

Hydrangea paniculata 'Magical Moonlight', early green stage.

beautifully shaped florets of a nice cream when fresh, which do not change colour. It also provides a good succession of flowers. *Hydrangea paniculata* **'Pink Lady'** is another plant suitable for the smaller garden, achieving some 1.75m (5½ft) by about as much across. The sterile florets are large, colouring quite quickly to an attractive pink, which develops around the margins of the cream florets. The hard-pruned plant in the trial produced large impressive panicles.

Of the more recent introductions due to be included in the new trial by the RHS in their newly developed Bridgewater Garden in the North West is *H. paniculata* **'Sundae Fraise'**, a seedling raised by M. Jean Renault of France, who at the time of writing has named and marketed some nine attractive plants, including the popular cultivars 'Vanille Fraise', 'Diamant Rouge', 'Dentelle de Gorron' and 'Diamantino'. 'Sundae Fraise' is compact, little more than 1.5m (5ft) tall, and holds its colour well into the autumn before fading. It opens with dense, roundish panicles, creamy white turning to a uniform pink, missing the rather dramatic bicolour effect of 'Vanille Fraise' at its best but nonetheless creating a charming picture. This promises to be an excellent cultivar for the smaller garden.

Hydrangea paniculata **'Diamant Rouge'** is another full head of sterile florets, transitioning to pink and red with time. A short stubby, full panicle, held well up by strong stiff shoots. One of the best red forms with a good succession.

Hydrangea paniculata **'Magical Moonlight'** is in the same zone as 'Limelight', the closely packed flowers opening with a greenish hue, turning white and ageing to a pink tone. Light pruning is recommended to ensure the heavy heads are held high. The 'Magical' Series was initially selected for suitability as cut flowers, and this is no exception, the flowers are held on long straight stems.

Hydrangea paniculata **'Little Lime'** is like a dwarf mounded form of 'Limelight', with small florets on a rounded inflorescence, suitable for the smallest garden. Flowers open lime green and are slow to turn white, before eventually taking on a hint of pink. Its exceptionally dwarf habit makes it useful in any small space or narrow border.

Hydrangea paniculata **'Wim's Red'** has a spreading, but modest habit, 1.5 × 2m (5 × 6ft), more wide than high. The influence of the dwarf-growing 'Dharuma', one of its parents, is evident. Red stems are topped by medium-sized informal rounded panicles, set in groups of fertile flowers. The ray flowers start white, gradually turn pink and finally achieve a uniform deep red. It was named for the late Wim Rutten of Holland.

Hydrangea paniculata 'Athena', close up, just opening (left) and with hosta (right).

Hydrangea paniculata **'Athena'** is a rarely seen cultivar from the de Belders and named for one of their grandchildren. It is a distinct cultivar in flower and catches the eye, with visitors to White House Farm always asking, 'What is that?' – a good sign of approval. White/cream sterile flowers have four petals, each about 3cm (1in) long, narrowly elliptic and well-spaced, giving an almost starlike effect on a loose, more open, rounded panicle, well-furnished with a good balance of fertile and sterile flowers. Flowers stay fully white before fading to green. With medium/light pruning it makes a free-flowering shrub with an informal presence of about 1.5 × 2m (5 × 6ft).

Hydrangea paniculata **'Le Vasterival'** (syn. 'Greatstar') has larger flowers in the same style as Athena but on a coarser plant, strong and vigorous and even in the open, not sufficiently free flowering, with a poor ratio of flower to foliage.

Hydrangea paniculata 'Praecox' makes a small tree.

Hydrangea paniculata 'Dharuma', early flowering, with a typical flattened corymbose panicle, late in its season.

Hydrangea paniculata **'Praecox'** is a very old variety, originating in the USA and said to have been introduced in 1893 at the Arnold Arboretum in Boston by Charles Sargent from seed collected in Hokkaido in northern Japan. It makes a spreading, rather stiffly branched small tree of 3–4m (9–12ft), producing its rather small, rounded white corymbs with toothed ray flowers on the old wood in June. Flowering on old wood of the previous year is significant, because if winter-pruned like other cultivars (which flower on growth of the current year), the flowering wood is pruned away and it will not produce flowers. It has a fragrance which Michael Howarth-Booth describes as a mix of 'horse and Chanel No. 5.'

By contrast, *H. paniculata* **'Dharuma'** is a small shrub, not much more than a metre tall, with the same flat corymbiform panicle with a loose cluster of white ray flowers, which quite soon turn pink. Some regard it as a possible hybrid with the closely related *H. heteromalla* but, although this cross was successfully made by Emile Mouillere in France in the 1930s, there is no evidence to support the assumption. It is of some garden merit but surpassed in cultivation by some of its seedling progeny. It has been used successfully as a parent in producing other early flowering forms and hybrids with a similar flattened corymbose panicle shape of inflorescence, such as the Dutch Bulk nursery at Boskoop's introduction of **'Early Sensation'** AGM, of modest proportions as a shrub of around 1.5 × 2m (5 × 6ft) and with white, shortish panicles borne at about mid-June on reddish stems. The flowers gradually turn a very pleasant pink and these contrast nicely with a crop of later borne fresh white flowers. It has an admirably long season and is a good all round cultivar.

These descriptions are, of course, only scratching the surface of the myriad cultivars now available, but those described are still among the best and most reliable performers. 'New' varieties are coming on the market almost continuously as I write and many are not yet widely available or tested in garden conditions against those already in cultivation. There is no requirement in registering a new variety to specify in what respects it represents an improvement or change on what has gone before, and I suspect the trend will be to produce bigger, fatter, redder flower-heads. The results of the current RHS trial at Bridgewater will give interesting pointers to any cultivars showing a notable improvement in garden performance compared to earlier cultivars.

Finally, in the meantime I have given a 'kennel' name to a cultivar raised from seed collected in Hunan province in China, towards the western limits of *H. paniculata* distribution. It is in complete contrast to the big bosomy mophead cultivars, having a light, open, airy, wraith-like panicle. The quite large ray flowers are held on long, slender thread-like pedicels and will move and flutter like butterflies in the slightest breeze. It is a very different cultivar, as yet unpruned, and it will be interesting to see the flowers on a hard-pruned specimen.

Two seed-raised specimens collected in Japan with similar flattened panicles. They both combine this character with early flowering and it seems there is a group within the species with these two distinct characters.

Hydrangea paniculata that flutters: a very different proposition (in first flower, top, and autumn colour, bottom).

CHAPTER 5

ASPERAE: ENDLESS VARIABILITY

There are three species in subsection Asperae (McClintock): *H. aspera* from the Himalaya and China and two Japanese endemics, *H. involucrata* and *H. sikokiana*. But recent revisions, once again, have split out species from those sunk into *H. aspera* in the McClintock monograph, discussed below.

The complexities of taxonomy

On a botanical trip into the mountains north of Baoxing in Sichuan in 1998, it had been raining more or less non-stop for two days, and drivers of four-wheel-drive vehicles had pronounced the steep tracks too vulnerable to rockfall and landslip to make it worth their while to take us any higher – plus it was still raining. So we slogged uphill on foot in the rain on a wide track to try to make more altitude, spurred on by the prospect of finding *Emmenopteris henryi* and *Carpinus fangiana*, only to be passed by two heavily laden lorries heading up to marble quarries high above. Having thumbed a lift on a third lorry, we discovered that the sacks we were sprawling on contained not rice but high explosive, so when the radiator boiled over and the driver put out his cigarette and took off to find a stream, we thanked him and bailed out, having gained at least some useful altitude.

At about 1,900m (6,000ft) we took a trackside break. I used the time to collect leaves from four plants of *H. aspera* growing where we were sitting. Though taken from comparable positions on each plant, and growing in the same track-side scrub within a radius of about 10m (30ft), all four were different in shape, size and vestiture (the type of hairs on the leaf reverse and shoot). Shape, size and vestiture were the three key characters used by McClintock to differentiate the four subspecies of *H. aspera* – differences not based on floral characteristics.

The four subspecies McClintock recognized were *H. aspera* ssp. *aspera*, *H. aspera* ssp. *strigosa*, *H. aspera* ssp. *robusta* and *H. aspera* ssp. *sargentiana*. The last two are now broadly recognised as species in their own right. *Strigosa*, in my view, is best treated as a highly variable (morphological) group, denoted by strigose vestiture, rather than a subspecies. My four adjacent leaf specimens in the wild did not neatly fit her analysis, as no two were quite the same. There were strigose and villous hairs, and what looked like intermediates. Leaving aside the difficulty of McClintock using only dried herbarium material as the basis of her study, it seemed to me that we were looking at a very variable species – difficult to pigeonhole botanically from its leaves. This was an assumption confirmed by later looking at many other living specimens, and by the fact that she had lumped no less than 31 taxa previously classified as discrete entities into the single species *aspera*. The flowers, though varying in colour, size and form, that is horticulturally, could not be separated botanically.

OPPOSITE PAGE: Two White House Farm aspera seedlings with large flowers of the same cross – 'Hot Chocolate' × 'Titania' – in sharply contrasting colour and style of both foliage and flower.

Close-ups of *H. aspera* leaf reverse – (a) strigose hairs, flat appressed (like sandpaper); (b) villous, long, curled, woolly hairs (like fur); (c) strigose but less hairy; and (d) an intermediate form.

A few leaves from *H. aspera* in cultivation showing differences of size, shape and vestiture of leaf reverse.

Plants previously designated as species or subspecies by earlier botanists, such as *H. villosa* Rehder, *H. longipes* Franchet, *H. kawakamii* Hayata and *H. strigosa* Rehder, were all brought by McClintock into synonomy under *H. aspera*.

These McClintock subspecies have all recently been reinstated to full species level in *The Flora of China*. In addition, a new laboratory phylogenetic analysis at the *Institut National d'Horticulture et Paysage* at Angers in France published in the *Tree Genetics and Genomes Journal* (2010) agreed with the Chinese classification of *H. villosa* and *H. kawakamii* as two distinct species.

Asperae distribution and variants

With *H. aspera* the gardener is spoiled for choice, thanks to its almost infinite variability in foliage and flower. This variability is not surprising for a species with such an extensive geographical range: from the Himalaya, Burma, through western, southern and central China, to the islands of Taiwan, Sumatra and Java.

The epicentre of *H. aspera* is perhaps western China, where it becomes a local pioneer, a weed species colonising bare ground such as landslips, new road cuts and scoured riverbeds, with myriad seedlings that establish quickly with the help of the Indian monsoon. This provides ample moisture throughout the *H. aspera* growing season, from April to October. In Yunnan, these conditions, together with a benign mild dry winter, means *H. aspera* is common in the warm temperate zone of about 1,800 to 2,400m (5,760–7,680ft), though forms of *H. strigosa* can be found at lower altitudes.

Hydrangea aspera in the garden – characteristics and challenges

A function of its variable combinations of longitude and altitude in the wild, hardiness is also variable within *H. aspera*, so it is difficult to make any meaningful generalisations or to specify hardiness according to the usual zone formulae. Some forms have stood the test of time and are reliably hardy, certainly for most temperate gardens; but others, such as the *H. strigosa* group, are found in the wild at low altitude, leaf out precociously and so are susceptible to both early and late frost, and thus useless for general planting. By contrast, the Villosa group, from higher altitude, especially collections from its northern limits in Sichuan, have a shorter growing season and consequently are tougher and reliably hardy.

This variable hardiness makes the location of any garden of obvious importance. Considering where *H. aspera* might thrive, there is a huge difference between, for example, a benevolent western seaboard garden and a cold inland plot in the Midlands, or between shaded sheltered woodland and a sunny and windy suburban street. In potentially colder or more exposed situations, individual siting is vitally important and care should be taken with placement within the garden to avoid potential frost pockets at all costs. (Chapter 10 has more comment on this.)

Hydrangea aspera has been misunderstood as a plant that is only suitable for planting in hectares of woodland, or an otherwise highly-protected environment. They can, in fact, be sufficiently protected by companion planting in small suburban plots by neighbouring common, tough flowering shrubs, such as

As a pioneer plant *H. aspera* is frequent in western China: here, left, with Guan Kaiyun, then Director of the Kunming Botanic Garden; middle, at Baoxing in Sichuan; and right, a typical *aspera* environment in the Upper Salween in Yunnan.

philadelphus, forsythia, flowering currant, cherries, crabapples or magnolias. For optimum performance, aim for some shade during the hottest part of the day and for reasonably moist soil during the growing season. It is important to provide wind shelter from the north and east especially: I found that *H. aspera* varieties that comfortably withstood a minimum of −10°C in still conditions in 2012 were twig-killed down to two years' growth level by −3°C in 2017 by an exceptionally searing easterly wind that blew for days.

In cultivation, time of flowering is also variable, and *H. aspera* forms and cultivars may be found in flower for several months from June to late October. This flexibility is an invaluable asset, especially for companion planting. For example, the cultivar *H. aspera* **'Farall'** peaks in mid-June, while the **'Taiwan'** AM form of *H. aspera kawakamii* is usually at its best in the second half of September and into early October.

Generous length of flowering time is also a valuable characteristic of all *H. aspera*, with most forms still playing a role in the garden landscape well into autumn, even after the flowers have lost their initial freshness. Asperas tend to age well, with flowers changing and fading but remaining effective in a background landscape of diminishing colour as the year wears on.

Another asset of *H. aspera* is variability in stature and habit, with forms to suit any size and quality of garden, from tidy suburban plot to extensive woodland estate. For example, *H. aspera* 'Farall' after twenty years is about 2 × 2m (6 × 6ft), while 'Taiwan' is some 3 × 7m (9 × 21ft) across. In the wild, there are arborescent specimens in warmer areas of more than 10m (30ft) in height and, at the same time, some forms of *H. strigosa* little more than 1m (3ft) tall.

Hydrangea aspera is not fastidious as to soils, provided they are well drained and not stagnant in winter. The soil pH is not a significant factor for healthy growth and, indeed, the species will succeed on chalk, given good provision of compost, leaf mould or bark mixed in at planting, and an occasional good mulch. Flower colour is said in the literature to be unaffected by pH, but I have reservations about this. The ray flowers may not change, but the central fertile flowers may be blue or close to blue on low

Early flowering *H. aspera* 'Farall', known for flowering all at once.

Hydrangea aspera 'Farall' can also have neat, compact corymbs (*see* later in this chapter).

pH acid soils. I have given the kennel name **'Blue Haven'** to one seedling with quite small ray flowers, but startlingly bright-blue fertile flowers. I am also growing a seedling from Arunachal Pradesh, with unusual suborbicular leaves and stunning bright-red young growth that in acid woodland flowered for the first time in 2022 with almost royal blue fertile flowers and large, white ray flowers (both discussed and illustrated later in this chapter). In the late Dr Jimmy Smart's garden at Marwood Hill in North Devon there was an impressive *H. sargentiana* in acid conditions with deep-blue fertile flowers, which when transferred to a neutral soil in Kent disappointingly reverted to the more typical washy mauve.

Pruning *H. aspera* is not necessary, except to remove old flower shoots for appearance's sake and for old plants, to take out old and dead wood from time to time.

Finally, the primary consideration for the gardener, *H. aspera* offers both flower and foliage, with a wonderful variety of colour, size and form.

Flowers lack the full-figured glamour of the mophead hortensias, being more understated, but are often elegantly and timelessly beautiful. The foliage plays a key part as a foil to the

Hydrangea aspera exceeding 10m (33ft) in Yunnan, western China.

A variety of ray flower forms of *H. aspera*, randomly picked from the garden at White House Farm.

Hydrangea sargentiana with bright-blue fertile flowers at Marwood Hill.

flowers. As Professor Michael Dirr wrote in his exuberant style: 'Leaves – wow… warm, fuzzy, like bunny ears, dark green above, variably pubescent, yet below grayish and cloaked in velvet, waiting to be stroked.'

Recent developments at White House Farm of coloured foliage forms with strong, pink flowers have extended the range, with upper leaf surfaces made darker and a leaf reverse ranging from dark magenta to warm pink, a dramatic complement to large pink flowers (*see* later in this chapter).

This all adds up to *H. aspera* offering a remarkable diversity of flower, form and foliage, which combined with a satisfactorily wide tolerance of growing conditions and with appropriate siting, should suit the range of preferences and purposes of most gardeners.

Hydrangea aspera cultivars

Perhaps the best known species and forms, because they are the oldest in cultivation, are *H. aspera* **sargentiana** and *H. aspera* **'Macrophylla'** AGM. The former was introduced by

Hydrangea sargentiana trichomes – characteristic fleshy hairs, defining the species.

Ernest Wilson in 1908 from a single location, and I believe has been reintroduced only once since then, by a Belgian PhD student at Ghent university (whose thesis was on *H. aspera)*, on a research expedition with Chinese colleagues following in the footsteps of Wilson in Hubei province. It is readily distinguished by the size of its leaves and the fleshy tufts of hair (trichomes) on its shoots, not unlike a moss rose, but without the fragrance. This is a totally consistent feature and firmly separates it from the other forms of *aspera*; and I believe qualifies it to be classified at full species level.

Hydrangea sargentiana is not a good plant for general planting, however: fastidious, hard to please and rarely seen in optimum condition, often presenting as a few gaunt and bony sticks dressed with poor foliage, even in woodland shade, where it frequently struggles with dry soil through tree root competition. The flowers are also modest, the white ray florets small relative to the foliage and the fertile flowers usually a dull greyish mauve. However, at its best, in ideal conditions, with good overhead light, shade for much of the day and a good supply of moisture at the root, it can create a dramatic effect, with strikingly large leaves on sterile shoots up to

Hydrangea sargentiana in flower.

Hydrangea aspera 'Macrophylla'.

47 × 32cm (18 × 12in) with a 20cm (8in) petiole. It flowers in July/August.

By contrast, *H. aspera* **'Macrophylla'** AGM is a good garden plant, hardy and less demanding in its cultural requirements, though like all asperas, it enjoys some shade and moisture during the growing season. The foliage is bold, but not as large as *H. sargentiana*, dark green and thus a good foil for the dozen or so whitish ray flowers, fringing a central dome of bright-purple fertile flowers. It is a striking shrub of some 2 × 3m (6 × 9ft), flowering over a good period in July and August. It is an easy shrub to please, and has been cultivated in the UK for many years, but it is not as widely planted as its quality warrants.

Another Wilson introduction, perhaps the best known *H. aspera* cultivar, is the form known as 'Villosa', which, because of its variability from seed, is now known as the 'Villosa group' in the RHS Plant Finder. As mentioned earlier, this has been recently recognised again as a species. The difference in value of its various seedling forms is wide, but the cultivar sold by Hilliers for many years as 'Villosa' has sensibly been given a clonal name of *H. aspera* **'Velvet and Lace'** AGM, and gardeners can do no better than acquire this superior form of an outstanding garden shrub. It is fully hardy, with narrowly ovate leaves felted below, and it is laced with flower in July and August. The ray flowers

Hydrangea aspera 'Velvet and Lace' in autumn colour (late September).

The late-flowering *H. kawakamii* group 'Taiwan' at its best in mid-September (syn. 'Taiwan Pink').

Hydrangea aspera 'Anthony Bullivant' – twice as wide as high.

are a pretty fresh mauve/pink, irregularly nodding around a dome of purple/blue fertile flowers, remaining effective in the garden into October, well after the fresh colours have faded to their pastel versions. It usually grows to about 2m (6ft) in height, spreading to perhaps 3m (9ft) across. It is not choosy about soils, and its long season makes it an excellent candidate for a partly shady corner of an herbaceous border. It is in the front rank of effective flowering shrubs.

A large, spreading form, lumped by McClintock into *H. aspera* ssp. *aspera*, but today often listed as a species, ***H. kawakamii*** Hayata, is also a very desirable and distinctive plant. The first introduction of recent years was by the late Maurice Mason from seed supplied by the Taiwan Forestry authority. I showed this plant to the RHS Woody Plant Committee where it received an Award of Merit as a plant for exhibition. It required a cultivar name, so after deep thought I named it *H. aspera* **'Taiwan'**. 'Taiwan Pink' is a synonym. 'Taiwan' is not a plant for the small garden, spreading after fifteen years, as someone once put it of a weeping willow, 'to about ten minutes across'. Mine is 7m (21ft) across and counting, and 2.5m (8ft) high. It is particularly valuable for being at its best in late September, when other flower colour is scant. It is also attractive in spring with new growth tinted with bronze, and a prominent white-felted underside to the leaves. The ovate/lanceolate summer foliage is a dark sea-green. The pale-pink ray flowers are quite small, on relatively long pedicels but complement the fertile flowers, which have conspicuous dark-pink buds opening to violet/purple flowers, the three elements making an attractive combination. For many years it has proved hardy in Kent, but has suffered superficial damage from once-in-a-decade prolonged cold winds. Bleddyn and Sue Wynn-Jones, to whom White House Farm is indebted for many fine hydrangea introductions, have brought in other excellent forms from Taiwan, including two they have named *H. aspera* 'August Abundance' and *H. aspera* 'September Splendour'.

Hydrangea aspera **'Anthony Bullivant'** is also ultimately a large plant. At Stourton House in Wiltshire, I measured the original shrub at 6m (18ft) across by 3.5m (11ft) high, named by the late Mrs Elisabeth Bullivant to commemorate her husband, and originating in their garden. It is one of my favourite forms, as the large, pink ray flowers that appear in July are almost orbicular in outline, the petals overlapping to form a virtually circular flower, making it particularly attractive. The foliage is ovate, acuminate and mid-green with a villous indumentum. It is reliably hardy.

Hydrangea aspera 'Sam MacDonald' showing its spreading habit.

Hydrangea aspera 'Peter Chappell', a true and quite compact albino.

Hydrangea aspera **'Sam MacDonald'** and *H. aspera* **'Spinners'** are two other reliable selections, the former with nice lanceolate, felted leaves, a lower habit than most asperas and mauve/pink ray flowers. It is a little less hardy than other varieties.

Three first-class plants for the smaller garden are *H. aspera* 'Peter Chappell', *H. aspera* 'Mauvette' and *H. aspera* 'Farall'. After ten years plus, none are likely to much exceed 2 × 2m (6 × 6ft) and all are usefully complementary in terms of different flower colour and flowering timing.

Hydrangea aspera **'Peter Chappell'** is a true albino, with pure-white ray flowers and pale-pink budded fertile flowers, opening white. The inflorescence is loose in outline with the ray flowers dispersed randomly around the corymb, giving the flower a touch of informality. In full light the foliage is also relatively small, some 15 × 5cm (6 × 2in) and in good scale with the plant. It will comfortably take more sun than most and is thus ideal for a mixed border in the smaller garden. It is a distinctive, attractive and versatile plant. A single plant growing in his greenhouse was given to me by the late Peter Chappell on condition that I propagated it. Enquiries failed to uncover its origin, so I named it to commemorate one of the UK's most celebrated and much loved plantsmen/nurserymen.

Hydrangea aspera **'Mauvette'** was raised in Holland in the 1980s and is similar to 'Peter Chappell' in every particular, except flower colour. As the name suggests, this is a pale lavender-mauve, a different and distinctive soft colour.

Hydrangea aspera **'Farall'** is little known and only recently appeared in the RHS Plant Finder. It is notable for its early flowering and its subtle colour, which has been regarded as 'refined', whether or not this is taken as a compliment, and might be best described as a wash of smoky mauve/pink. This colours an exceptionally free-flowering symmetrical shrub, which in full light is clothed to the ground, with all shoots flowering simultaneously to create a compelling picture. On the other hand, there is little continuity of flower. The foliage is an attractive mid-green, tinted and fringed with bronze. It is valuable for flowering early for an aspera, in June. The plant was given to me by the late Michael Haworth-Booth many years ago: I named it to commemorate the Haworth-Booth garden. Its species' affiliation is unclear but it is clearly in the Asperae section.

Hydrangea aspera **'Titania'** is a most dramatic form for flower. Introduced by Tom Hudson at Tregrehan in Cornwall, it is a plant with large leaves, prominent red petioles and huge, fringed, white ray flowers on strong shoots on young plants. The dark-green elongated heart-shaped leaves have a pale-green reverse, and can reach 32 × 22cm (12 × 8in) in size, with a long, dark-red, 17cm (7in) petiole and red veining on vigorous,

Hydrangea aspera 'Mauvette', a well-named Dutch selection.

Hydrangea aspera 'Farall' – shapely, floriferous and early.

Hydrangea aspera 'Titania' is a true Titan with large ray flowers – a montage of *aspera* forms with 'Titania' at the centre.

The fimbriation of 'Titania' corymbs is striking.

sterile shoots that become smaller, broadly ovate leaves on flowered shoots. The foliage provides an excellent backdrop for its stunning, very large, white fimbriate ray flowers, up to 8cm (3in) in diameter. My plant was damaged by a bout of freak winter cold in Kent ten years ago, but bounced back vigorously, and is likely to succeed when well-sited, and become a much sought-after plant, especially for milder gardens. Seedlings do not match the fimbriation of the original.

Hydrangea ***longipes*** has distinctive foliage, with heart-shaped or broadly ovate, rough, papery leaves up to 22cm (8in) long by 20cm (7in) across and prominent petioles, slender and notably long, almost as long as the leaves. McClintock places it in *aspera,* sinking it into *H. robusta*. I have seen the same plant under these two different names, and believe it warrants full species definition – as it had before McClintock's revision of the genus. *Hydrangea longipes* is not a plant for the small garden, growing to 2–2.5m (6ft) and with a lax habit, sprawling for up to 5m (15ft), elegantly sweeping the ground like a large crinoline. Both fertile and ray flowers are white, the latter conspicuous and up to 4cm (1½in) across on a corymb of some 15cm (6in). The best flowering form I have seen was at Trelissick garden in Cornwall many years ago, with ray flowers up to 5cm (2in) in diameter. The then curator kindly gave us cuttings and we have distributed material to other interested gardens and think that the large flower size qualifies it to be given the clonal name of **'Trelissick'**. It is an excellent plant for thin woodland where there is plenty of light. Where it has room to spread freely as a specimen it is a magnificent plant. It also flowers early, in June. There are also new wild introductions of this species from the Crug Farm nursery.

A plant with foliage of similar size and shape but with a different chartaceous texture and indumentum, conspicuous bright-red buds, pink/mauve fertile flowers (in *H. longipes* the fertile flowers are white) and white ray flowers is an interesting

Hydrangea longipes 'Trelissick' spreads elegantly with its long petioles.

The Trelissick form of *H. longipes* ray flowers (centre) compared with typical *H. longipes* (left and right).

Hydrangea robusta showing conspicuous bright red buds (left) and in full flower (right).

A low shrubby form in the *strigosa* group hanging over the river at Ya An in Sichuan with willow-like leaves and loose, mauve flower corymbs, conforming to Wilson's var. *angustifolia*.

Hydrangea aspera ssp. *strigosa*: Wilson's var. *macrophylla* flowering in late September.

Hydrangea aspera ssp. *strigosa* with conspicuous white birch-like bark.

recent introduction, and I believe is true *H. robusta*. I saw a single plant in fruit north of Baoxing in Sichuan and thought it distinctive, even in drab, wet cold and almost dark, autumn conditions, and having grown it on from seed thought it may be new to cultivation until later seeing an identical old plant at Gresgarth, Lady Arabella Lennox-Boyd's magical garden in Lancashire. This was labelled 'Villosa' and a gift from Kew. It makes a big, rounded bush some 3 × 3m (9 × 9ft). I have not traced the original Franchet description of *H. robusta*, which might throw further light on it. It is a striking plant in bud.

The *Hydrangea strigosa* group is a highly variable group. Indeed, it is difficult to generalise as to its character, as it appears at its two extremes both as a small, narrow-leaved shrub, rather like a diminutive weeping willow flushed with mauve when in flower, and as a massive arborescent multi-stemmed bush or small tree with large, ovate leaves.

Ernest Wilson characterised *H. strigosa* thus: 'In size and shape of flowers the Chinese plant seems exceedingly variable, and the following forms pass gradually into each other – var.

macrophylla, var. *sinica* and var. *angustifolia.'* Clearly, plenty of scope for variety there. It is a low-altitude plant, so in cultivation generally not suitable for cold gardens. It leafs out early and most forms flower extremely late, some well into October, when the flowers often lack sufficient warmth to develop fully. Some forms are first class, but only in favoured gardens, free of late frosts and hard winters. On the more arborescent forms, bark can be exfoliated or bound and a conspicuous white, to the point where I irresponsibly characterise a 4.5m (14ft) specimen at White House Farm pruned up to a tree as *H. strigosa* forma *betuloides*.

Hydrangea aspera forms with dark-purple and pink-coloured foliage

There has been a recent influx of coloured foliage seedlings in Europe, the USA and Japan. These all originated in China, where such forms are not infrequently encountered. Their dark foliage generally has a distinctive red/purple leaf reverse and

Recently developed dark-leaved forms of *H. aspera* at White House Farm: expanding the range of foliage, flower and form.

Two White House fam aspera crosses showing the variety possible; pink rayflowers continue to be effective even when mature and reversed.

Hydrangea aspera 'Koki' in flower.

creates a unique effect in the garden landscape. Flower colour varies from off-white to rich pink.

At White House Farm I have been crossing these dark-leaved forms for about fifteen years, and have raised some new cultivars that have proved to be hardy, have retained their striking foliage and improved in flower colour, with strong, large, ray-flowered pink variants giving excellent value in the garden. It extends the range of planting opportunities by a quantum amount.

Some of the crosses had green foliage but with large, white ray flowers: these can be as much as 7cm (3in) across.

Hydrangea aspera 'Gongshan' with late sun coming through the foliage.

Late sun shining through the plant can bring these darker foliage colours alive and create spectacular effects of maroon, orange and bronze, making some varieties valuable as contrasting and colourful foliage plants, with the flower extra icing on the cake.

One of the boldest of these dark foliage forms, with affinities to *H. strigosa*, has large, dark-green, broadly ovate leaves with a maroon-red reverse and pink flowers. This has been named *H. aspera* **'Gongshan'** after the town on the Upper Salween. Such coloured-leaf forms, which occur occasionally in the wild, do seem to generate a fair proportion of seedlings true to type, but many revert to the normal green.

Hydrangea aspera 'Gongshan' has proved not to be reliably frost-hardy. In one exceptional cold spell shoots were killed back to three-year-old wood. However, this had the same effect as hard pruning, and regrowth was dramatic, with exceptionally large leaves, some 32 × 16cm (12 × 6in) on shoots of 1.2 to 1.5m (3½–5ft), with a rich-red reverse, vividly illuminated by a low sun and perhaps suggesting that *H. aspera* 'Gongshan' is worth growing, hard-pruned annually, purely as a foliage plant. It is an excellent choice for warmer, relatively frost-free gardens.

Hydrangea aspera **'Koki'**, a selection from a Mikanori Ogisu introduction, has proved satisfactorily hardy in Europe. Its purple young foliage matures to dark green with a red reverse, dark shading and bronze/purple petioles. Its large leaves are broadly ovate, and twice as long as broad, providing a perfect backdrop to the whitish ray flowers surrounding a dome of purplish-pink fertile flowers that emerge in late August and continue through September. A bushy, spreading plant, it has come through testing winters literally with flying colours. In good conditions – part-shade and with plenty of moisture at the root – it makes an impressive large dome, some 3m (9ft) high by 4m (12ft) across. It is called both 'Burgundy Bliss' and 'Plum Passion' in the USA, with the usual obligatory alliteration.

Hydrangea aspera 'Koki', a big bold plant with a big bold leaf that turns from purple to green as it matures.

An unnamed White House Farm *H. aspera* seedling with crimped flowers and a pink leaf reverse.

An unnamed form introduced from seed collected in Hubei province in China by the same PhD student at the University of Ghent in Belgium, who reintroduced *H. sargentiana*, has turned out to be hardy at White House Farm in Kent, in spite of its strigose vestiture. It makes a medium-sized bush of neat, upright but open habit some 2.5 × 2.5m (8 × 8ft). The foliage is lanceolate to ovate/lanceolate, tapering elegantly to a long, acuminate tip with a notable leaf reverse of colourful pinkish-mauve. The ray flowers are slightly crimped, with pointed sepals and a pale-cream pink. The bark is a feature, an exfoliating creamy white. It is an attractive and conspicuous plant in the garden landscape.

White House Farm *Hydrangea aspera* crosses and seedlings

I have been breeding new dark-leaved forms of *H. aspera*, using various combinations, with the objective of achieving a moderate grower that is fully hardy and has a bold clear pink flower along with a bold dark leaf. A second objective, with a subsequent series of crosses, has been to raise a hardy arborescent form that could be pruned up on to a single stem, to allow the purple or pink foliage to be comfortably viewed from below. *Hydrangea aspera* leaves that are surface green but with maroon undersides are particularly striking from that perspective. With the sun coming through them they vary from a rich, dark orange to an eye-catching orange-bronze-purple.

My own seedling selection, *H. aspera* **'Hot Chocolate'** was launched in 2014 by the Jan Oprins' nursery in Belgium on the European market, and is now widely available in garden centres. It was an open-pollinated seedling from Koki and the flower difference is clear.

It has these dual-coloured leaves, ovate and usually three times as long as wide, dark green above and a maroon-red below but when young, suffused with purple. Perhaps even more significant, gently cupped ray flowers of good size and a soft, clear pale-pink are presented in this seductive foliage setting. Some five to ten of these pink ray flowers are held informally around corymbs of bluish/purple fertile flowers, creating an informal bicolour and domed inflorescence. The habit is bushy, of moderate height and spreading, about 1.5 × 1.5m (5 × 5ft). It flowers in July. A sister seedling of 'Hot Chocolate' after a decade has reached under a metre in height, and is a genuine compact form of *H.* aspera with a smaller elegant leaf, perfect for small gardens or pots.

A typical WHF dark-leaved aspera hybrid, of garden merit as a foliage plant even without the pink flowers. Both pictures were taken at the same time on a cloudy day: the bronze foliage is the result of semi-transparency and a green leaf surface combined with pink reverse.

Flowers from *H. aspera* 'Hot Chocolate' (left) and 'Koki' (right).

Hydrangea aspera 'Hot Chocolate' (top) and its dwarf sister (below).

My late wife had a favourite seedling, which I have named *H. aspera* **'Rosemary Foster'** in her memory. In 2019, it was awarded a Gold Medal at a Belgian trade show, and is now available in the UK. It makes a dense, relatively compact plant with a spreading habit, with dark-green foliage and a leaf reverse of very dark reddish-purple. The leaves are narrowly ovate or lanceolate with a long, elegant tip, which gives the plant a lightness of touch and which provides the perfect foil for the flowers of three colours – white ray flowers, with fertile flowers that are dark red in bud and bright purple when fully open. It flowers late, in August, and the white ray flowers eventually turn reddish, the colour of Victorian plush. In its autumn colour it is an effective flowering shrub into November and beyond if the weather is favourable.

I crossed *H. aspera* 'Rosemary Foster' with *H. aspera* 'Hot Chocolate' to attempt to create a race of late-flowering coloured-leaved plants with good, richly coloured flowers. I grew on some eight promising seedlings currently being trialled. One of these, named for my Canadian rosarian friend Darlene Sanders, has a spectacular tri-coloured flower, with a few white ray flowers surrounding a great mass of fertile flowers, bright red in bud, with a prominent brush of bright-blue stamens on opening. The plant is untypically drawn up in a trial situation at present so ultimate size and habit is as yet unclear, but it is a striking plant in flower and likely to have a moderate spreading habit in a more open position.

I also crossed *H. aspera* 'Hot Chocolate' with *H. aspera* 'Titania', which has large, deckle-edged ray-flowers, with the aim of producing a plant of relatively modest growth, with dark-red foliage and very large fimbriate, perhaps pink ray flowers. Fimbriation and size were the main objectives. The majority of seedlings have inherited the *H. aspera* 'Hot Chocolate' dark foliage and most do have unusually large, sumptuous ray flowers, of a very bold, clear – in some cases hot or lipstick – pink.

Hydrangea aspera 'Rosemary Foster' – in bud (left); at its peak of flower in mid-August (middle); and in close-up with leaf reverse (right).

Hydrangea aspera ('Rosemary Foster' × 'Hot Chocolate') in bud.

Hydrangea aspera ('Rosemary Foster' × 'Hot Chocolate') in full flower.

White House Farm *H. aspera* seedling ('Titania' × 'Hot Chocolate') with large lipstick-pink ray flowers showing some fimbriation.

However, the fimbriation was usually lost, evidently a recessive character: but these new plants have all inherited the vigour and impressive foliage and something of the flower size of *H. aspera* 'Titania', and are being trialled in various conditions at the time of writing.

Mophead forms of *Hydrangea aspera*

Ernest Wilson introduced a form of *H. strigosa* in 1904 (W 4902) from Mount Omei in Sichuan with all sterile ray flowers, in effect a 'mophead' form of *H. strigosa*. This has been lost to cultivation, presumably killed off by subsequent cold winters. But almost a hundred years later, American plantsman/explorer Dan Hinkley found a similar globose-headed plant on Mount Omei and successfully introduced it, naming it *H. aspera* **'Elegant Sound Pavilion'** after a nearby temple of that name. It flowers in late September and early October and appears not to be frost-hardy in Kent, with two small plants killed in successive winters. The immature flowers are pale green maturing to white, then once more reverting to green as they age, a delight to flower arrangers. When growing strongly in shelter, it constantly reverts to lacecap flowers, so with me it has not been a first-choice plant.

Another unusual *H. aspera* White House Farm seedling comes close to a mophead type, with crowds of relatively small ray flowers invading the whole inflorescence. We gave

Hydrangea aspera 'Elegant Sound Pavilion'.

it the initial kennel name of '72 Flowers', this being the number of small ray flowers we counted on a single typical big flower-head. This plant has proved perfectly hardy here, and was admired by a visiting French nurseryman and his wife, so we named it **'Aurelie'** in her honour. I believe it is now being propagated in France. It is quite neat in growth, perhaps 2.5 × 2m (7 × 8ft) across.

An unnamed sister-seedling of *H. aspera* 'Hot Chocolate' caused a minor sensation when it flowered as a genuine mophead, with similar reddish foliage to its siblings. Though with every appearance of being a runt, I kept it because it

White House Farm's semi-mophead form *H. aspera* 'Aurelie', with its large scatter of ray flowers.

Unnamed sister-seedling of *H. aspera* 'Hot Chocolate' with a mophead flower.

was interestingly weak and took some time to get going; when it flowered it proved to be a unique colour combination: the crowded flowers are a suffusion of claret, greenish cream and pink. It is a compact grower and now flowering freely, so is full of promise. We are trialling it in various positions to test its stamina, hardiness, sun tolerance and shade-bearing capabilities.

In the meantime other 'mophead' forms of *aspera* have been introduced by US collectors; and at a fascinating slide presentation to the RHS Woody Plant Committee of some of his new plants, Mikinori Ogisu showed a beautifully coloured, fully globose specimen, which he assured us was hardy. It looked a very exciting new find. As far as I am aware it is not yet in cultivation in the UK.

There is a White House Farm strain of true blue *aspera* seedlings that have proved hardy and undamaged here in Kent over some fifteen years. One we have given the 'kennel' name of *H. aspera* **'Blue Haven'**. It has dark-green, modestly sized foliage and strongly domed, clear blue fertile flowers with white blue-veined ray flowers. The ray flowers are both smaller and fewer

Two new blue *aspera* seedlings – 'Blue Haven' (top) and a promising unnamed companion (below).

than ideal, but nonetheless, the bright blue makes this an attractive plant. It is currently growing on pH 5.5/6.0 and requires trialling in neutral soil to test the potential impact on colour of a higher pH.

I have given a kennel name to a selection made from a range of plants raised from seed collected in Arunachal Pradesh, N. India, donated from two different sources. It is an interesting new form with suborbicular leaves on long petioles, furnishing what to date is a plant of relatively modest size, bushy, with a spread wider than high. The name refers to the young growth, which is a deep rich red compared to a mere bronze tint on other seedlings. The ray flowers are white, arranged randomly among a large, very loose tumble of bright blue fertile flowers. It has plenty of promise as an effective garden plant, but in its early days in assessing hardiness.

Hydrangea involucrata

The second horticulturally important species in the Asperae subsection is *H. involucrata*. The name describes the characteristic ball-like spherical bud, in which the flower is literally 'wrapped up' (from the Latin word *involucrum*, a wrapping, cover or envelope) by a series of bracts.

The Japanese name for *H. involucrata* is *Tama–ajisai*. This translates as 'globe hydrangea', which like the Latin species' name, also refers to the large, round flower bud, wrapped in an involucre, a whorl of bracts enclosing the entire inflorescence. This bud is a distinctive and defining feature, found only in

White House Farm *H. aspera* seedling from Arunachal Pradesh with red spring growth (left) and with its blue and white inflorescence (right).

Hydrangea involucrata: all forms have the characteristic spherical bud.

H. involucrata in the Asiatic Asperae subsection and in some hydrangeas in the Cornidia subsection, the latter a race of tree-climbing, evergreen hydrangea species found mainly in Central and South America (*see* Chapter 9).

The type species of *Hydrangea involucrata* in the wild

Hydrangea involucrata is a Japanese endemic, growing mostly in woodland and at relatively low altitude, from around sea level up to about 1,500m (4,800ft). It was first collected by von Siebold and described by him in 1829. A related species with involucrate buds, *H. longifolia* (for some taxonomists classified as a subspecies of *H. involucrata*) is found outside Japan, in Taiwan, where it is endemic.

The type species is usually a small, rather open shrub of about a metre by as much across, though significant variations in habit occur, from small plants of less than 1m (3ft) to shrubs that will reach as much as 4m (12ft) in a warm maritime situation. The ovate leaves vary in size, from 10cm (4in) to as much as 25cm (10in) long, and about twice as long as wide. Both upper and lower surfaces are covered with pale adpressed hairs, as are many other parts of the plant, notably the shoots and pedicels.

The fertile flowers are typically a bright lavender/mauve blue, with some four to six white or very pale lavender ray flowers 2–3cm (1in) across, often attractively waved, which form a pleasant contrast and invariably turn a shade of green as they mature. The type flowers from mid-July into August. Double and mophead forms have been discovered in Japan and some of these are highly effective in the garden landscape until well into November.

Typical wild form of *H. involucrata* in Japan.

Hydrangea involucrata forms in cultivation

The *H. involucrata* species remains relatively rare in cultivation, with historically a reputation for a lack of hardiness and a weak constitution. Howarth-Booth writes that it is often seen at only 0.6m (2ft) high or less 'owing to all growth above ground being winter killed'. Bean also reports that it is often killed back more or less in winter, but will still manage to flower on new shoots. In fact, this 'remontant' character, of flowering on the new growth of the current year, gives the species a significant and unusual advantage in frost-prone areas, because if twigs are winter-killed, it will still produce flowers that same year from new shoots springing like a herbaceous plant from the base. However, it does have the disadvantage of flowering only terminally. This means that it is not as showy or profuse in flower as, for example, most *H. macrophylla* forms, which often also flower from axillary buds down the stem. However, some of the newer double mophead *H. involucrata* forms can now rival these in sheer weight of flower.

Recent experience of hardiness has also been rather better, perhaps due to an increasingly consistent run of relatively mild winters or, possibly, to the fact that previous specimens of *H. involucrata* in commerce were propagated from limited stock that may have been less hardy than more recent introductions. These have now proved to overwinter satisfactorily, building up into spreading shrubs, with some cultivars even succeeding in colder continental gardens. I have grown all those described for more than twenty years in Kent and they have performed wonderfully well in the garden without damage.

Optimum growing conditions for *H. involucrata* are the same as those for *H. aspera*: good overhead light, some shade during the hottest part of the day and moisture available during the growing season. As with all hydrangeas, frost pockets are to be avoided and shelter from cold winds is essential. Some

Hydrangea involucrata Viridescens – freely flowering in near total shade.

cultivars will happily thrive in almost full sun, provided there is moisture at the root, with tall sheltering companion plantings helping to maintain a degree of humidity.

However, full sun throughout the day and wind exposure can lead to some leaf burn at midsummer. Flower colour in some double *H. involucrata* cultivars is also influenced by the sun/shade balance, with deep shade producing pale – in some cases, white – flowers as compared to colourful pinks, buffs and greens in full light. Having said that, some forms are among the best of all shade-bearing plants in continuing to flower unusually freely in a gloomy situation that never sees the sun (in particular *H. involucrata* **'Viridescens'**). A key garden value of the cultivars is their lateness: fresh blooms bringing an air of spring to an English September, sometimes into October.

The leaves of *H. involucrata* have dense, flattened hairs on both surfaces and like a leafy Velcro, will collect spent flowers and debris from taller companion plants unless removed by gently shaking them free. In wet weather these can cling, coagulate and leave ugly brown patches on the foliage that can be unsightly throughout the year.

No pruning is required, apart from removing dead wood and old flower-heads in late winter in the interests of tidiness. As with most hydrangeas, propagation is simple, with half-ripe summer cuttings under polythene or mist rooting easily (for more on this, *see* Chapter 11).

Effective *Hydrangea involucrata* garden forms recently introduced

In the past century, cultivated forms of *H. involucrata* were limited to the type species and one small double form introduced to the UK in 1906, known as **'Hortensis'**. In recent years, Japanese botanists and collectors have discovered a variety of mutations in the wild, some with very attractive and distinctive double flowers, far removed from the form of the type species both in flower and habit, and valuable additions to our garden flora. Mr Takeomi Yamamoto features some fifteen different *involucrata* cultivars in his *Colour Guide to Japanese Hydrangeas*, translated into French through the agency of Corinne Mallet of the Association Shamrock at Varangeville, home to the French national collection of hydrangeas. A number of these have recently been introduced to Europe and are now becoming generally available.

In addition to these Japanese selections, new cultivars of *H. involucrata* have originated in Europe, all essentially with the same general floral characters as the type. The late Wim Rutten in Holland raised many hydrangeas, among which two cultivars of *H. involucrata* have been named. *Hydrangea involucrata* **'Late Love'** is now available in Europe. It is a compact plant, notable for its long four-month flowering season from July to October. It is a particularly healthy form, and said to be less susceptible to spider mite under glass. The flowers are typical of the species, with six to eight shapely white ray flowers in an even circle, 2.5–3cm (1in) across, each with four tepals, fringing a relatively flat corymb of pale blue fertile flowers. It continues to throw odd flowers deep into the autumn.

Hydrangea involucrata **'Blue Bunny'** is a protected cultivar in the USA. I have no personal experience of this plant, but it appears to be another typical form of the type species, with perhaps a more generous number of ray flowers. It is marketed *inter alia* on the basis that it has the particular capacity to flower in sun or half-shade for a long season on new wood and, thus, like most *H. involucrata* cultivars, will give a return, even if cut by frost. The origin of the name is not known – apart from the obligatory alliteration, it may be that the white ray flowers are thought to resemble a rabbit's bobtail, although most woody gardeners would certainly not welcome that sight.

Hydrangea involucrata **'Oshima'** is a form collected by Corinne Mallet on the island of that name in a maritime environment in

Hydrangea involucrata 'Late Love', very similar to the type species.

the south-east of Japan, where the species will grow as tall as 4m (12ft). It is vigorous and surprisingly hardy, flowering in August, with six to eight white ray flowers 2–3cm (1in) across, each with four sepals. The ray flowers are slightly nodding, on long pedicels, arranged around a corymb of lavender-blue fertile flowers, typical of the species. This form is recognised at subspecies level by Japanese botanist Hayashi as *H. involucrata* ssp. *izuensis*, which thrives in the benign climate of the Izu peninsular and islands on the east coast. Plants of this type – with exceptional height, and a thicker and more hairy leaf – are also known in Japan as *Raseita–Tama*. They are valuable in favourable conditions of moisture and shade where there is a good spot for a more substantial specimen with a similar flower to the type.

One of the very best involucratas that resemble the type species is named *H. involucrata* **'Chichibu'** after the place where it was found in the wild. It was introduced by the University of Ghent Botanic Garden, where there are excellent collections of many *Hydrangea* species. It forms a neat domed bush, after eight years about 0.75 by 1.5m (2 × 5ft) across, with dark-green foliage setting off domed corymbs densely packed with bright lavender-blue fertile flowers and ringed informally with some six to fifteen (eighteen) white ray flowers. These are 2–3cm (1in) across, with five sepals that are lightly and most attractively waved and fluted. They eventually droop nicely, becoming bright green. When happy, *H. involucrata* 'Chichibu' will sucker gently without becoming invasive; and has borne some sharp spells of winter cold without damage, so appears to be hardy. It is a very attractive form now available for general planting.

A very distinctive cultivar, particularly valued by flower arrangers for its vivid green and mauve/purple flowers, is *H. involucrata* **'Viridescens'** AGM. Curiously, this is known by the anglicised name of 'Green Tama' in Japan. It was found by a Japanese nurseryman near Mt Fuji. It is a compact plant, usually some 1.5m (5ft) tall by 1.5m (5ft) across. The fertile flowers are a dull mauve in bud, opening into a mass of small lavender flowers with a white calyx, in evidence after petal fall. They form an attractive contrast with the six to eight (max. fifteen) ray flowers, which are a consistent bright green with a tiny lavender flower at the centre. These are 1.5–2.5cm (½–1in) across with four to five (eight) sepals. It is exceptionally tolerant of shade, not too much drawn up out of character under trees and flowering on every shoot, even those of the current year, even when never in direct sunlight. I play a game with visiting groups and say that anyone can take a cutting if they are able to find a shoot without a flower bud. Visitors who know plants go straight for the lower interior of the plant, only to find that even short shoots of that year at the base bear flower buds, despite being confined to the gloom. When grown in good fertile soil, the terminal inflorescence is sometimes supplemented by flowering shoots from the upper leaf axils, usually with fewer (four to five) ray flowers in the corymb. Its garden value is confirmed by its RHS Award of Garden Merit (AGM).

Of the double forms, *H. involucrata* **'Hortensis'** has been established in cultivation for many years. It was described as a Japanese garden plant in 1867, but not introduced into the UK until 1906. Opinions vary as to hardiness, with Bean finding it not completely hardy at Kew. However, I have grown it in Kent in a sheltered corner for more than thirty years without damage, with larger plants behind forcing it to sprawl forward on to a brick path, where it revels in the shelter, sun and warmth, and

Hydrangea involucrata 'Chichibu', a charming form named for the region in Japan where it was discovered.

Hydrangea involucrata 'Viridescens', not only flowers freely in deep shade but also on shoots of the current year.

Hydrangea involucrata 'Hortensis' is a double form that revels in a warm situation.

flowers freely. The curious flowers are a muddled mass of doubled petalage in an untidy corymb, with the outer flowers like tiny roses, 2–2.5cm (1in) across, fringing a loose agglomeration of doubled flowers of various sizes in the interior. These are halfway in size between fertile and ray flowers, buff cream, enlarging to a clear rose pink and fading to a greenish cream suffused with pink. This rather tortured description belies the overall effect, which is very attractive – although on a relatively small scale, as growth is limited to about half a metre high and one metre across. The individual flowers open progressively over a long period, with others fading prettily to create an effective display for many weeks from July onwards, into late September. Sunshine is needed to create this miscellany of colour: in deep shade the corymbs are less full, the flowers pure white, and in partial shade colours appear but are more muted. It is a plant that repays careful siting both in terms of its cultivation and its effect.

Hydrangea involucrata 'Plena' is pretty at close quarters.

A more recent introduction with an altogether tidier inflorescence is *H. involucrata* **'Plena'**. The name derives from the six to twelve (max. twenty) double ray flowers, each of which have some seven to eight sepals in double rows, the inner row about half the size of the outer and pointing forward. These are white, faintly washed with lavender and poised upright on relatively long 2cm (1in) pedicels. When mature, they stand pert and proud of the pinkish lavender fertile flowers. The plant grows to around 1.5 × 1.5m (5 × 5ft). It makes no great impact of colour in the garden landscape as a whole, but is very pretty at close quarters and ideal for the front of a border flanking a path. It will take a good deal of sun if supplied with some moisture at the root.

A better choice for significant impact in the garden is *H. involucrata* **'Yohraku'** AGM, also discovered on an island in

Hydrangea aspera 'Yohraku': a large bush with a long season of double flowers, that increase to as many as eight flower layers in a chain by end of season.

the Izu archipelago. It is softly colourful in a medley of pastel pinks, creams and greens, changing colour week by week as the flowers mature into autumn. It begins to flower at the end of July and peaks around the end of August, while continuing to open a few fresh flowers well into October. The crowded ray flowers are numerous, double, with a central row of smaller tepals creating a shapely little flower some 2.5–4cm (1–1½in) across, not unlike a tiny rose. They are heaped informally around a loose central area of mauve-pink fertile flowers. The individual ray florets gradually mature to green. At the same time, a further row of pink sepals grows from the centre of each of the original ray flowers, after they have turned green. This process continues over several weeks to form up to eight rows of further tepals along an extended pedicel, the ultimate row a mauve pink. As autumn fades into winter, these form a loose ruff of pendant green, creating a unique effect as they project like tight daisy chains from the corymb (*see* Chapter 6). *Hydrangea involucrata* 'Yohraku' makes a bigger shrub than most involucratas, but nonetheless neat and fully clothed to the ground, some 2 × 3m (6 × 9ft). In a sheltered situation it is perfectly hardy, but is also happy in a relatively sunny open position, as long as it has some shelter. I have grown specimens in Kent in shade, half-shade and sun for some twenty years without frost damage, even in the testing winter of 2013. It is an unusual and attractive hydrangea that well deserves its AGM and should be recommended for more general planting.

Also gardenworthy and easy to grow is *H. involucrata* **'Sterilis'**, known as 'Temari-Tama' in Japan. It has been in cultivation here for many years, but for reasons that are entirely obscure, is seldom seen in gardens. Its ball-shaped flowers appear in the second-half of July and continue into August, with soft green buds maturing to a creamy white, faintly flushed with pink, the inflorescence enlivened by a few mauve-pink fertile flowers scattered randomly throughout, as well as at the heart of each individual floret. The bush rarely exceeds 1 × 1.5m (3 × 5ft) and is ideal for a small garden. Though its style could be described as a 'mophead', it has a feminine lightness, delicacy of flower and subtlety of colour that belies that image and is matched by few other hydrangeas of this type for charm. It is reliably hardy.

Hydrangea involucrata 'Sterilis' is good for the small garden.

Another cultivar similar in style and habit to *H. involucrata* 'Sterilis' is *H. involucrata* **'Handemari'**, discovered by Mr Kiyoshi Yamaguchi in the Shizuoka prefecture. The name means 'partly double'. It differs from 'Sterilis' in its looser inflorescence, with each ray flower more faintly washed with pink, and with many more mauve-pink fertile flowers in evidence. It too rarely exceeds 1 × 1.5m (3 × 5ft) across, starts to flower in mid-July and goes on deep into August before fading to a quiet green. A wonderful garden plant.

Hydrangea involucrata **'Mihara–Kokonoe'**, like *H. involucrata* 'Oshima' above, was also found on the island of Oshima in Japan, and to be seen at its best it needs ideal siting in mild conditions, preferably with plenty of humidity. Like most of the double forms, it is not easy to describe as it changes colour week by week, so any photograph or description is a mere snapshot in time. It is a strongly double form, which begins as a central mass of green buds ringed by 2–3cm (1in) of white florets poised like small butterflies. These gradually fade to pale green as the central agglomeration explodes into a mass of double flowers of different sizes, mostly clear white washed with green, creating a dense large ball of subtle cool colour. It is a relatively frost-tender 'soft' plant that takes some time to come into leaf, and in Kent does not flower until late September and early October. It needs a slow, warm autumn to flower well. In wet, cold weather the flowers will not develop fully and will spoil. The flowerheads overall are heavy and tend to hang, especially when wet. If not cut by frost, it can reach 1.5m (5ft) in height. It is a dramatic flowering plant, but can be recommended for only the mildest areas where it will need plenty of sun to perform well.

Thought to be a sport from this 'Mihara–Kokonoe' variety is a relative newcomer called *H. involucrata* **'Toraku–Tama'**, that Robert Mallet of the celebrated Shamrock *Hydrangea* collection at Varengeville-sur-mer near Dieppe describes as a 'magnificent variety'. It received 'le Merite de Courson' in 2009. Robert describes it as a vigorous shrub 1.5 × 2m (5 × 6ft) that flowers at the end of August, continuing until frosts begin. Large inflorescences of pure white fade to green, with soft foliage like *H. aspera*. Habit, size and foliage appear to be similar to *H. involucrata* 'Yohraku'.

The University of Ghent Botanic Garden has introduced another promising *involucrata* cultivar, 'Hanabi-Tama', which, while not dissimilar from other double forms of the species, looks to have a clearer lace-cap form, with pinkish-mauve fertile flowers crowned by a ring of star-shaped double ray flowers with a yellowish centre, on long pink pedicels.

Hydrangea involucrata 'Handemari' is similar to 'Sterilis', but with more fertile flowers in a looser head.

Hydrangea involucrata 'Mihara-Kokonoe' needs warmth to be at its best.

Finally, closely related to *H. involucrata* but endemic to Taiwan, ***H. longifolia*** has the same conspicuous round flower bud, but bears elegant, narrowly ovate to lanceolate leaves, deep green above, with a greyish reverse. McClintock sank it as a subspecies of *H. involucrata*, but most authorities restore it to species level. According to Wei and Bartholemew in *The Flora of China*, the distinctive two-branched hairs of *H. longifolia* easily technically differentiate it from *involucrata* as a species. Horticulturally, *H. longifolia* also differs in not being fully frost-hardy and slower to flower, opening usually in the second-half of September or October, with the latest flower buds, therefore, sometimes failing to develop in an English climate. In a very mild winter and a favoured location it is semi- or even fully evergreen. It usually bears three to four white ray flowers, nicely poised, with ovate to obovate sepals 2–2.5cm (1in) long and deep-pink fertile flowers in a relatively small, loose corymb. It is an elegant plant, but likely to succeed in only mild gardens, such as Tregrehan in Cornwall, where it is splendid.

While forms of *H. involucrata* are still not widely planted in gardens, it is notable that two cultivars were awarded the AGM in the 2013 revision, encouraging recognition of the subsection's garden value. These are not simply plants for favoured woodland gardens on the coast; I grow them here in White House Farm at 150m (500ft) in Kent, with the right aspect and careful siting, and they have been a joy for more than twenty years. They are valuable plants for the ordinary, average garden and do not warrant their earlier reputation for winter vulnerability, particularly in light of what is becoming our norm of milder winters. There are now about a dozen nurseries in the UK offering some form of the type species of *H. involucrata*. In addition, an almost full range of the double forms is available both here and in Europe: highly distinctive hydrangeas in both form and colour. Thoughtfully placed to get the best out of them, they add a lot of character and interest as well as colour to the late summer and early autumn garden.

The final and horticulturally least important species in subsection Asperae is ***H. sikokiana***. A Japanese endemic, it is related to the American *H. quercifolia* as the only species within the genus with technically pinnately lobed leaves. The lobing on *H. sikokiana* is variable, usually more toothed than lobed and often confined to the upper-half of the leaves, which are very hairy and up to 20cm (8in) long. It makes a shrub up to about 2m (6ft). The flowers consist of a wide, rather flat corymb of fertile creamy white flowers hung with one or two inconspicuous ray flowers. It is a fastidious plant, lacking a strong constitution and of little horticultural merit. Introduced by the Crug Farm team, it remains a rare collector's item.

Hydrangea involucrata 'Hanabi-Tama'.

Hydrangea longifolia – related to *H. involucrata*, endemic to Taiwan.

Hydrangea sikokiana, showing flower, and leaf shape.

CHAPTER 6

MACROPHYLLAE: MAINSTAY OF THE SUMMER GARDEN

McClintock recognized *H. macrophylla* as the only species in this subsection, with four subspecies. Two of these, *H. serrata* (pictured opposite) and *H. stylosa*, are today generally accepted at species level, the latter the only member of the subsection to be found outside Japan, with a wide distribution along the eastern Himalaya into Burma, China and Vietnam.

The Macrophyllae: colour range and change

As I explained in Chapter 1, the historical introductions of various forms and hybrids of the *H. macrophylla* subsection, and their subsequent development and exploitation, have made these hydrangeas the most popular and widespread garden hydrangeas of the twentieth century. To most, the word 'hydrangea' means a brightly coloured mophead form of *H. macrophylla* – the flower style also called 'Hortensia' – albeit with expected variations of colour. The fashion for these plants since their emergence on the scene in the nineteenth and early twentieth centuries has been essentially a European affair, matched latterly by the USA and, to a degree, Japan, its natural home. But the history of hortensias since then has been one of continuing growth, with a veritable explosion of new cultivars over the last 30 years.

The Japanese Hortensia is among the most variable, unstable and unpredictable group of plants in the genus *Hydrangea*. To begin with, there is its capricious tendency to change colour depending on differing soil conditions, famously turning blue in acid and pink in alkaline soils, which then gives rise to its inclination to show every intermediate shade between the two. Then there is its remarkable ability to also change colour during the season – the flower in June is not the same as the flower in September. Both of these colour-changing habits are universally recognised and in Japan regarded as a rather mystical transformation, with all kinds of sentiments attached to it, such as a sign of prosperity or a change of heart, or simply the inevitable passage of time. But we should add to this a genetic instability, leading to mutations or sports of flower and foliage in nature that yields anomalous forms, different shapes, colour behaviours and foliage to flower relationships – a whole series of unstable qualities.

As a botanical subsection of the genus, *Macrophyllae* comprise the plants commonly understood as 'macrophyllas' – that is, the mopheads or hortensias so widespread and familiar as cultivars – and *H. serrata*, an important species also in the subsection. These two contrasting but phylogenetically compatible species give great potential for creative breeding – on which more later. As it is, the wide variety of mophead *H. macrophylla* forms has been rather less developed in Japan than in Europe

OPPOSITE PAGE: A White House Farm *H. serrata* seedling that is a nice bright red in neutral soil. It transitions to dark blue in acid conditions. Serratas are in the same subsection as macrophyllas ('Macrophyllae') and deserve greater recognition.

Hydrangea macrophylla 'Mme Emile Mouillere' showing three flowers of different ages. This is a natural photograph, not an arranged confection. A fresh flower (left), autumn colour (centre) and a mature flower (right). Taken in October.

Hydrangea macrophylla 'Ayesha' (*see* later in this chapter), originally a sport, the central blue flower partially reverting to the original form (left) and *H.* macrophylla 'Mikore Chidore' with some ray flowers reverting to their original larger size (*see* later in this chapter).

Hydrangea serrata in the wild in Japan, in *Fagus crenata* forest, Tanbara swamp, Gunma Prefecture.

and the USA; but the opposite is the case with *H. serrata*, prized for millennia in Japan for its understated delicacy of form and subtle variations in colour and style.

All hydrangeas in Japan have been appreciated for centuries and deeply embedded in Japanese culture, with references in the very oldest poems going back as far as the eighth century. Potted *H. macrophylla* plants are commonly given as presents on Mother's Day, and even traditional sweets are provided in the shape of hydrangeas during the flowering season. But *H. serrata* is nationally revered. Many towns claim it as their official flower, with parks and temples holding annual hydrangea festivals where enthusiasts exhibit their prized forms of *H. serrata*. The delicate beauty of *H. serrata* is thought to be perfect for the simple performance of the tea ceremonies during the season, and to help lift the spirits during the summer rains, when it looks its best.

Hortensias: the most popular Hydrangea of all

Today the changeable, unstable characteristics of *H. macrophylla* provide the great variety of form and colour that make it the summer plant of choice wherever it can be successfully grown, right across the world. Its chameleon character is at the heart of its popularity; it is the mainstay of the summer garden. In Europe and the USA, the Hortensia or mophead *H. macrophylla* is the ubiquitous hydrangea of the garden centre, supermarket and bargain store, found on sale everywhere in the southern hemisphere at Christmas; swathes of it add blocks of colour to agricultural country shows and fairs, not to mention summer sporting events like Wimbledon. It is everywhere across Europe, anchoring village churches in Brittany in billowing masses of colour, gracing town centres, beautifying restaurant fronts across France, Belgium and

Reliable and decorative *H. macrophylla*, often used in pots *en masse* to beautify public events.

Switzerland, and ramping along the southern Cornish coast of England, where old, neglected plants take the salt winds in their stride and overflow every cottage garden. In the USA, which has taken a major role in developing new hydrangea cultivars in the 21st century, it has been the subject of vigorous market promotion, and lies behind the designation of hydrangeas as the 'plant of the millennium'. Macrophyllas are a mass phenomenon in Europe and the USA, selling in their hundreds of thousands; in the USA, hydrangeas are now second only to roses in sales.

A contrast in cultures or a function of markets?

It is no coincidence that the vast majority of names attached to *H. macrophylla* cultivars are European in origin rather than Japanese; by the same token, the overwhelming variety of names for the delicate charm of the related *H. serrata*, integrated into Japanese culture for millennia, are Japanese rather than European, and *H. serrata* is gaining ground but has yet to make its full mark in Europe. A good example of this is that in his 2002 *Colour Guide to Japanese Hydrangeas*, the authoritative Mr Takeomi Yamamoto lists only 67 *H. macrophylla* cultivars ('gaku ajisai' in Japanese) of which a mere sixteen are mophead hortensias, the remainder mostly lacecaps akin to the simpler wild species. By contrast he catalogues no less than

Takeomi Yamamoto's 2002 *Colour Guide to Japanese Hydrangeas*.

172 *H. serratas* ('yama ajisai'), including the subspecies *H. yezoensis* ('yezo ajisai').

Having said that, it has to be admitted that the big mophead *H. macrophylla* cultivars have become more popular in Japan in recent years, perhaps as they have become more universally developed, and worldwide demand for them has grown.

One could speculate that this different emphasis represents a difference in taste of two very different cultures: a European inclination for large, self-proclaiming masses of mophead colour without much regard for form, versus a Japanese taste for the subtle and delicate detail of shape and texture, rather than simply colour. I am far from qualified to take any view about Japanese taste, but I have noticed in Japan how small treasures in pots are revered, in particular coloured-foliage forms and variegation, as well as anomalous forms of common plants, such as those with double flowers. A 'garden' in urban environments where the front door gives straight on to the street could consist merely of a large window box with azaleas, peonies and camellias, flanked by a wisteria in a pot; to be replaced, when they have finished flowering, with roses, hydrangeas and other summer flowers, keeping the show going through the year. In urban Japanese gardens there seems to be great pride in a perfect display of both colour and form in pots, contained and deliberate, with care and attention to detail.

One could also argue that the European focus on the hortensias is largely a function of the nature of horticultural markets. Almost all the first hydrangea introductions from Japan to

H. macrophylla 'Madame Emil Mouillere' (1903) showing early flowers (left) and *H. macrophylla* 'Générale Vicomtesse de Vibraye' (1906) (right).

Europe were hortensias. In raising new varieties from seed, European nurserymen like Lemoine, who successfully grew his business by breeding many fine cultivars of *inter alia* lilac, deutzia, weigela and philadelphus, spotted the potential of the hortensia in combining flower power and colour with suitability for culture in pots. The first few big blue mophead flowers went down particularly well with the public. This should not be taken to imply that market-driven nurserymen lack the passion for plants encountered by the amateur. They can equally love their plants, but they do run businesses and always need to have an eye on the bottom line: the consuming public will buy what it has already seen or heard of, so once a tradition is established, it is likely to be sustained. There is no question that expansion of the pot-plant trade drove a trend, set a standard among aware nurserymen and created a market. Mouillere plants like the big white *H. macrophylla* 'Madame Emile Mouillere' in 1903, or the butterfly blue *H. macrophylla* 'Générale Vicomtesse de Vibraye' in 1906 are still near the top in the popularity lists today over a century later, a real test of and tribute to their quality.

The rise of *Hydrangea macrophylla* cultivars, or 'What's in a name?'

Long or unattractive names can disadvantage the distribution of a plant: branding has always been important. The right name is key in forging an image for both sales and identification purposes: rose innovator David Austin, for example, has been particularly good at this, creating a bespoke nomenclature as part of a cohesive marketing campaign for his roses around the world. This has provided a really strong base for the competitive establishment of his brand.

By contrast, some of the rather overdone hydrangea names of the early twentieth century – *H. macrophylla* 'La Petite Sainte Therèse de l Enfant Jesus', for example – are not easy on an anglophone ear, any more than *H. serrata* 'Miyama No Murasaki'. Many European hydrangeas are named to commemorate anonymous people, friends and relatives of the raiser. Research into the names alone could read like a journey through European social history. In conversation once with the late Christopher Lloyd about raisers of hardy hybrid rhododendrons, which had their heyday in the UK during the same era as early hydrangeas, I offered the view that breeders got around the convention of never naming plants you have raised after yourself by naming them for their wives. This was of course at the end of a Victorian period when wives were treated as an adjunct to their husbands, so being addressed as Mrs Fred Smith was quite normal. It killed two birds with one stone, allowing the main man to pay tribute to himself while still honouring the lady. But when we got on to naming roses and hydrangeas and I expressed my delight in wonderful names like 'Mme Sancy de Parabere' or 'Mme Lauriol de Barney' or 'Mme Faustin Travouillon' that conjured images of a beautiful, immaculately dressed and coiffed *mère de famille* at table for Sunday

Hydrangea macrophylla 'Mariesii Perfecta' syn. 'Bluewave': this is a rather overrated plant, slow to 'blue', fading in sun and usually seen as a mauve/grey amalgam of washed colour.

lunch with her dutiful children, while the father, donning white gloves, carved the guinea fowl, Christopher, who had an impishly ironic sense of humour, accused me of being a romantic and said they were more likely to be the wives of some provincial nurseryman, wrinkled with age after a gruelling lifetime of hard work. He later sent me a photo of one such – a portrait of a frowsty old dame with a severe expression, face lined by years of toil and dressed all in black.

Names of obscure people, being non-descriptive, are harder to remember. This could account to some extent for those plants falling out of the mainstream and eventually disappearing. Haworth-Booth was in favour of renaming some of the more obscure, personally named hydrangeas, giving them brief, descriptive, memorable names. I'm not sure the rules of primacy in nomenclature allow this, but he justified changing *H. macrophylla* **'Mariesii Perfecta'** to **'Bluewave'** and *H. macrophylla* **'Mariesii Grandiflora'** to **'Whitewave'** (both pictured in Chapter 1) on the grounds that because these plants had nothing in common with the parent Mariesii (which Lemoine mistakenly thought was a distinct species, so had kept the link), it was quite proper to rename them. He also suggested changing the names of the best cultivars from the names of people to epithets like Scarlet Passion, Blue Dream or Mood Indigo. I'm not sure these colourful monikers are improvements, but as a nurseryman he supposed, probably correctly, that they would be better 'selling' names.

A forgotten English breeder

A discussion of hortensia names brings to mind the work of English breeder Henry James Jones, who has been the sole breeder of note of hydrangeas in the UK in the entire history of the genus. He ran his own nursery business in Hither Green, a suburb of London, in the early 1900s, coming upon hydrangeas

H. J. Jones

H. J. Jones wrote in his catalogue of 1920:

> A good bed of hydrangeas is one of the most beautiful sights it is possible to conceive; they are a surprise and a delight to all who have tried them, and their cost is small compared with the effect and enjoyment they give. Everyone who has planted them on my recommendation has given them unstinted praise and has been more than satisfied in all ways. I always advise planting mixed colours as the most effective. *There are many shades, all of which blend beautifully and give a most wonderful effect and give that delightful feeling of Hydrangea madness, which I possess in a large degree.* I wish everyone who wants something new and beautiful in the garden would take my advice (though interested) and plant a few. [My italics]

And in an extract from *The Horticultural Advertiser*, 1 June, 1927, it says:

> Hydrangeas. – The way this plant has come to the front in recent years is wonderful, and the splendid stuff shown was good evidence of its popularity. Mr. H. J. Jones, who was the pioneer introducer of the newer sorts, again made a magnificent show, including all the latest novelties, several of them of his own raising, grown in the highest style.

H. J. Jones won two gold medals for his hydrangea displays at Chelsea, in 1926 and 1927. In all he raised and named 26 cultivars, all hortensias, 23 of which were named after people he knew or admired when he was active in the RHS. These included one 'Miss' and seven 'Mrs' (with husband's initials!), King George, Lord Lambourne, the Duchess of York, Princess Marina and the then very young Princess Elisabeth. These upmarket names during the post-war social revolution of the roaring twenties suggest H. J. Jones was well-connected through his earlier herbaceous successes. Nine of Jones' hydrangeas were given the RHS Award of Merit (AM), an award for plant excellence 'for exhibition', i.e. in vases or pots, judged by RHS committee on the basis of quality of flower, foliage and presentation, rather than performance in the garden (hence the need for an additional RHS AGM or Award of Garden Merit).

Hydrangea macrophylla 'King George', a rarely seen H. J. Jones plant, flowering in the Hillier Arboretum.

towards the end of his life, having previously achieved international recognition for his work in breeding new chrysanthemums and other herbaceous plants. There is some information about him online, and a 2010 article about Mr H. J. Jones in Robert Mallet's *Journal of the Association Shamrock* by Roger Dinsdale, in French, who speculates that his interest in the genus may have been triggered by the accolade of a plant raised by Mouillere being named for Jones's wife – *H. macrophylla* 'Mrs H. J. Jones' – which also received an Award of Merit from the RHS in 1923. Or this could have simply been taken as friendly encouragement by the French nurseryman after Jones' first successful efforts. But by the time H. J. Jones had finished his late-in-life passion for breeding hortensias, he had raised and named 26 cultivars and won two gold medals at Chelsea for his hydrangea displays.

It is odd and somewhat sad that the achievements of H. J. Jones in hydrangeas do not appear to have been much referenced since. Few of his plants are still in cultivation in the UK; they are more likely to be found lurking in collections somewhere on the Continent. As a group, they would be a fascinating snapshot of horticultural fashion in that decade, and the social currents with which it reciprocally evolved. H. J. Jones' lone stand and his subsequent obscurity as a hydrangea breeder is an example of the puzzling lack of any significant tradition for growing and breeding hydrangeas in the British Isles. As the UK's only breeder representative, he deserves more recognition; certainly some of his plants would reward rediscovery. Hortensias are surprisingly long-lived, so this may not be a hopeless endeavour.

Hydrangea macrophylla in the wild

In Japan, distribution of *H. macrophylla* in the wild is confined to a relatively limited area on the coast south of Tokyo: namely the Izu peninsula and surrounding islands, plus the island of Oshima. It is thus essentially a maritime species, enjoying the conditions that go with that situation: a relative freedom from frost, high rainfall (175–250cm/year or 70–100in/year) and

Some of Corinne Mallet's collections of *H. macrophylla* in Japan demonstrating random variation in flower style.

humidity. Its name, meaning 'large leaf', reflects its rather thick, almost leathery, shiny green leaves, in general ranging from about 7 to 20cm (3–8in) long, and usually about one and a half times as broad as long, that give it good resistance to salt-laden winds. I say 'in general' because many variations of *H. macrophylla* leaf forms can be found in the wild, from almost round/suborbicular, to elliptic or with long points, or prominent marginal serrations. *Macrophylla* twigs are typically strong and stout, and often dotted with dark-brown lenticels, and some shoots from the base in optimum conditions can reach a metre in a year. In the wild, flowers are found mostly in simple lacecap form, but anomalous forms with different sized- and shaped-florets and mophead configurations are not infrequent.

The benign coastal situation of the wild species also means that *H. macrophylla* enjoys a rather long growing season, of perhaps eight or nine months; of which around half is spent either in bloom or with still attractive, faded colour.

Reliability and performance of *Hydrangea macrophylla* in gardens

This length of the growing season in a maritime, humid and frost-free environment makes the hortensia vulnerable in cultivation in growing seasons shortened by frost at either end of the growth period. Its natural long period of growth exceeds what in many climates is only a five- or six-month frost-free period, making *H. macrophylla* particularly susceptible to the dangers of frost in both late spring and early autumn. It often optimistically starts growth in English gardens in February, with buds rapidly pushing out as temperatures rise and daylight begins to lengthen, but also continues making buds and shoots as late as the end of November, which can lead to bud kill in early next-year's growth, or blackened and tip-killed unripened autumn shoots.

This bud-tenderness excepted, *H. macrophylla* is actually reasonably winter-hardy, certainly in the UK, if the shoots are able to ripen adequately the previous season to form hardened, mature, cold-resistant new growth. The balance of sun and shade has an

Hydrangea macrophylla buds can start pushing as early as late February.

impact on this hardening, with sun being necessary for adequate ripening. Having said that, *H. macrophylla* is often killed to the ground in very cold areas like Nova Scotia, where winter temperatures fall to around −20°C; explaining the historic engagement there with the potential of the fully hardy montane sections of the genus, such as *H. paniculata* and *H. serrata*.

In an open situation, a typical *H. macrophylla* will make a rounded, bun-shaped, densely foliaged shrub around 1–1.5m (3–5ft) high by about the same across. In shade it will obviously become more open, and might be drawn up to 2m (6ft) or more. I have seen an old, neglected, unpruned, shaded plant of *H. macrophylla* 'Ayesha' well over 3 × 3m (9 × 9ft) in the favourable climate of the Dandenong ranges in Victoria, Australia, covered spectacularly in its vivid blue small-cupped flowers in an acid soil.

The flowers of cultivated *H. macrophylla* plants, in both mophead and lacecap forms, can be found in an almost infinity of colour: white, cream, through pale cold-pink to warm rose to rosy red and deep red, and the equivalent of all these tones in blue. The darker the red, the deeper the blue (or in some cases purple) in acid soil conditions. There is little as striking as a good specimen of *H. macrophylla* massed in darkest blue, except perhaps when cold pinks slowly convert to a stunning, exquisitely pale blue. (*See* Chapter 10 for the soil factors that influence changing hydrangea flower colour and how this might be adjusted by soil treatments to suit individual taste.)

Hydrangea macrophylla 'Ayesha' in neutral soil.

The commercial development of *Hydrangea macrophylla* (lacecaps) – the Teller Series

Until the end of World War II, most of the hortensias planted were old trusted mophead cultivars produced either before World War I or in the 1920s and 1930s. But in the 1950s, the Swiss research establishment at Waedenswil began work on developing a series of lacecap cultivars to provide an alternative to the big mopheads in the pot plant trade, which by then had dominated the market for the previous 75 years. The flowers were close in character to the wild form and grown initially to compete as pot specimens with mopheads, not for the open garden. The Swiss named their plants the Teller Series: *Teller* being German for 'plate', a more prosaic allusion to the flatter, plate-like shape of these flowers than the English term 'lacecap'.

The Swiss crossed a *H. macrophylla* wild white lacecap form with an old red mophead called *H. macrophylla* 'Toedi'. The first results were all white and lacecap-type, and it was clear that both characters of the wild form were dominant. With further backcrossing, the result among the ensuing seedlings was 50/50 between lacecap and mophead forms. In total they eventually named 26 new cultivars, all but two named after birds, and in 1964 released them for propagation to a number of Swiss nurseries. When this had yielded sufficient numbers of young plants, they sent a full collection of the series to a number of different horticultural establishments in Europe to test their suitability for outdoor planting: Basel Botanic Garden, Brisago Island in Italy, Mainau in Germany, Arboretum Kalmthout in Belgium, Esveld in Boskoop in Holland and Exbury in Hampshire, UK.

I have not seen any published formal assessment of these tests to date, but the Teller Series are now widely grown in gardens, and my own experience in growing fifteen of the Teller cultivars over many years is wholly favourable. They are healthy, hardy, relatively compact plants with clear and clean colours in appropriate soils.

The trend the Tellers began for marketing 'Series'

The 'Tellers' pioneered a trend for marketing a 'series', a collection of hortensias comprising selected cultivars with a similar style, colour or set of characteristics that were often (though not always) the happy results of a single hybridising strategy. These were increasingly raised and named as series by European, Japanese and American breeders.

In the 1960s the 'Japanese Lady' Series was launched by Ebihard, followed by several more in the 1990s and early 2000s, notably: the 'City Line' Series by Rampp in 1990; the 'China City' Series by Kees in 1993; the 'Hovaria' Series by Hofstede in 1996; 'For Ever and Ever' by Klaveren in 2005–8; the 'Posy Bouquet' Series by Sakamoto' in 2000; 'You and Me' by Ichie in 2000; the 'Dutch Lady' Series by van der Spek in 2002; and the 'Magical' Series by Kolster in 2008.

From the USA have come the 'Endless Summer' and 'Royal Majestics' Series, both bred by Michael Dirr in 2001–2: the 'Double Delights' Series by Ball in 2013 and 'Lets Dance' by Wood, 2005–8.

This is not a comprehensive list and there are more appearing with great regularity. Some of the best examples of individual plants within these series are discussed in the cultivar descriptions in this chapter.

Two contrasting examples from the Dutch Ladies Series, 'Selma' (left) and 'Sheila' (right).

The Shamrock Collection at Varangeville-sur-Mer

Since the 1960s, there has been widespread development and hybridising of new *H. macrophylla* cultivars in the Netherlands, France, Japan, Belgium and the USA, accompanied by the assembling of impressively representative collections in gardens and parks that incorporate both newly bred cultivars and also plants of historical significance. Arguably, the best and most comprehensive of these hydrangea collections is the Shamrock collection in France, at Varengeville near Dieppe on the Channel coast, created over a period of some 40 years by Corinne and Robert Mallet. More than 1,800 cultivars are on display, representative of all subsections of the genus, but with a high proportion (approximately 1,000) of colourful *H. macrophylla* cultivars, for which Varengeville has a favourable climate. It includes many plants collected in the wild, as well as in gardens by Corinne Mallet over her many visits to Japan. It could easily be the biggest collection of hydrangeas in the world on a single site.

It is fully documented, and Corinne has published a series of encyclopaedic books listing and featuring images of many of the plants in the collection. In 1994, they created a Society of Friends of the Shamrock Collection, which compiled and published a valuable reference volume, the *Index of Cultivar Names*. This sets out a comprehensive list of hydrangea species and cultivar names, with dates, breeders and synonyms. The Society also publishes a regular well-illustrated and informative newsletter, available free in English on their website. The collection is open to the public and there is no better treasure house to visit in order not only to enjoy the beauty and interest of a globally-recognised collection, but also to check out your own preferences in making hydrangea planting choices for your own garden.

At the same time, new varieties continue to pour out of the USA, Europe and Japan, generating a bewildering number of cultivars of *H. macrophylla* from which to choose, both old and new. In the current edition of the RHS Plant Finder, there is a choice of some 350 hortensia cultivars. Many new varieties may not necessarily be an improvement in garden performance over the old and bold; it is perhaps advisable to keep some distance

Perspectives on the impressive Shamrock Collection at Varangeville.

from the marketing hype and assess the new not simply in terms of innovative flower shape, size and colour, but as good, reliable garden plants. There can be a significant and sometimes uncomfortable difference between a plant in a pot on the show bench, mollycoddled and fed and watered in a sheltered environment, and the same thing exposed to the vagaries of potentially brutal weather in the open garden. Tiptoeing through the endless numbers of cultivars, the question is how to sort out the best garden performers, as opposed to those selected for pot plant or cut flower sale in the florists' shop.

The RHS Hortensia trial and AGM cultivars

In the case of *H. paniculata*, which also suffers from a similar situation of oversupply, the results of the excellent RHS trial in 2008 can provide a reliable guide. An equivalent four-year trial of hortensias in 2014 was rather less satisfactory, mostly because of variation in on-site conditions. To get full value from a horticultural trial, it is essential to arrange every aspect of the growing environment with the closest attention to detail to ensure that every plant is subject to identical conditions. The only possible variable remaining should be the plant itself. Such consistent management of the plant environment enables effective and definitive cultivar performance comparison. Time and type of propagation, pot composts, growing-on in same-size pots, planting dates, pruning, aspect, climate, irrigation, soils, pH values, mulches, fertilisers and pest control – identical standards must apply throughout.

The trial was divided between two climatically different locations: one in a maritime, virtually frost-free environment, the other in an inland woodland. Some plants were in full sun, others in varying amounts of shade, and variable tree root competition meant there was different moisture content, even with irrigation. All these challenging variables meant that assessing relative hardiness, growth rates, quality of foliage, together with freedom, continuity, duration and colour of flower, was not totally straightforward. These less than ideal common conditions inevitably had some impact on assessment. The panel members sitting in judgment took this into account in finalising the awards and the end result still provides a useful guide for the gardener in making his or her own cultivar selections in outdoor conditions, and it remains a useful reference in short-listing good garden plants.

The RHS *Hydrangea macrophylla* AGM Trial Results

Of 172 trial entries the following 25 cultivars were selected for the award:

- *Hydrangea macrophylla* **'Trebah Silver'** is vigorous, upright with strong stems. A sport from *H. macrophylla* 'Ayesha' with slightly larger, flatter, long-lasting flowers of a soft pale-purple, opening from yellow/green buds.
- *Hydrangea macrophylla* **'Colorado'** is part of the Riverline Series. An open grower but compact, with dark-green foliage. Bright-red eye-catching flowers on strong stems, successional, with a long season.
- *Hydrangea macrophylla* **'Koria'** is noted for its flat, deckle-edged, strongly fimbriate white flowers of good size on a neat dense, compact bush. Fertile flowers blue or mauve/pink. Definite personality.
- *Hydrangea macrophylla* **'Pengwyn'** is an informal lacecap with white ray and fertile flowers, making ovate, pointed florets, held well above olive-green foliage. A good, strong, upright grower to about 1m (3ft). Nothing to do with arctic birds – named after a Cornish village.
- *Hydrangea macrophylla* **'Etoile Violette'** was introduced from Japan by Corinne Mallet at L'Association Shamrock in 1994. A strong grower to 1.5m (5ft) with open-spreading habit. Double lacecap flowers of purplish-pink to blue in acid conditions forming medium-sized flower-heads with pointed, starry florets.
- *Hydrangea macrophylla* **'Dancing Snow'** (synonym 'Wedding Gown') from Ichie, Japan in 2006. A white lacecap with attractively balanced, semi-double flowers on long pedicels scattered across the inflorescence, eventually maturing to a striking autumn red. Fertile flowers cream, making an attractive overall picture. Probably best in part-shade. One of my favourites.
- *Hydrangea macrophylla* **'Brugg'** from Haller, 1971. A good weatherproof, strong, red mophead on a compact, well-balanced plant to 0.5m (1½ft), packed with flowers. Best in a sheltered position with some shade.
- *Hydrangea macrophylla* **'Schoene Bautznerin'** from Dienermann, 1970: synonym 'Red Baron'. Sturdy, upright, large, crowded bright-purplish red/pink mopheads, well supported on thick stems. Weatherproof and flowering on young plants.
- *Hydrangea macrophylla* **'Dalian'** has dark-red flowers held well above good foliage on strong stems. Holds colour well in long season. Performed well throughout trial.
- *Hydrangea macrophylla* **'Together'** is part of the 2008 'You and Me Together' Series from Irie. In the USA the same plant is registered as 'Forever' and 'Ever Together'. Rated one of the best in the trial, it has a long succession of double, pink flowers in crowded mopheads held well above the foliage, becoming paler at maturity. Achieves a bright, clear blue in acid soils. Compact mounded habit.

The 'Magical' Series, bred by Peter Kolster and Cornelis Eveleens in the Netherlands from 2001 onwards, were originally developed for the cut-flower market but turned out in the trial to be hardy and well-suited to garden cultivation. The following were awarded the AGM:

Hydrangea macrophylla 'Trebah Silver'.

Hydrangea macrophylla 'Etoile Violette'.

Hydrangea macrophylla 'Colorado'.

Hydrangea macrophylla 'Koria'.

Hydrangea macrophylla 'Dancing Snow' syn. 'Wedding Gown', in bloom (above) and autumn colour (below).

Hydrangea macrophylla 'Pengwyn'.

Hydrangea macrophylla 'Brugg'.

Hydrangea macrophylla 'Schoene Bautznerin' syn. 'Red Baron'.

Hydrangea macrophylla 'Dalian'.

Hydrangea macrophylla 'Together'.

Hydrangea macrophylla 'Magical Colourdream'.

Hydrangea macrophylla 'Magical Revolution'.

Hydrangea macrophylla 'Magical Rhapsody'.

Hydrangea macrophylla 'Magical Wings'.

Hydrangea macrophylla 'Magical Garnet'.

Hydrangea macrophylla 'Magical Harmony'.

- *Hydrangea macrophylla* **'Magical Colourdream'** has a compact but open habit with mophead flowers that have strongly frilled, crowded red/pink florets with a light centre, maturing to pale pink. The frilled flowers are striking against foliage touched with bronze and a red edge.
- *Hydrangea macrophylla* **'Magical Revolution'** has a compact, upright, dense habit with pale lilac/purple, tightly crowded, bun-shaped mophead flowers, set off by olive-green foliage. Terminally flowering on every shoot.
- *Hydrangea macrophylla* **'Magical Rhapsody'** has a unique mophead flower of yellowish-green with an eye of palest mauve/purple and a fimbriate edge to gently cupped and thick-textured florets. Basically a green flower set in dark-green foliage on an upright but open, if not rather lax, shrub.
- *Hydrangea macrophylla* **'Magical Wings'** has an upright but open, bushy habit, with sun-proof white mophead flowers, with slightly fimbriate margins, first opening and later maturing to yellow/green.
- *Hydrangea macrophylla* **'Magical Garnet'** has strong red, crowded mopheads set in glossy olive-green foliage, free-flowering on a tidy, even bush.
- *Hydrangea macrophylla* **'Magical Harmony'** has a compact habit with pale olive-green foliage and strong stems that support pale purplish-pink mopheads, freely borne over a long season. Weatherproof and a consistently good performer.

The following cultivars were all reconfirmed as AGM plants following the 2014 trial:

- *Hydrangea macrophylla* **'Altona'**, bred by Schadendorf in 1931, with an AGM in 1992, is a compact 1.5m (5ft) vivid and deep-blue mophead in a low pH, that turns a remarkable brick red in the autumn. In alkaline conditions it is deep rose-pink. It enjoys part-shade, but will take some sun.
- *Hydrangea macrophylla* **'Générale Vicomtesse de Vibraye'**, bred by Mouillere in 1909, first awarded the AGM in 1992. This is often abbreviated in the trade to just 'Vibraye': a vigorous, spreading plant up to 2m (6ft), a wonderful clear blue in acid soil, blues easily but flowers freely all at once with little continuity, in crowded mophead corymbs; in alkaline conditions a refined pale pink. It is reliably hardy and will also take full sun, but gives better results in half-shade. With me this is a first-choice plant, unsurpassed. *Hydrangea macrophylla* **'Mousseline'** bred by Lemoine in 1909 is similar in colour and habit, but lacks lenticels on the shoots. Nonetheless that is another excellent plant, often confused with 'Vibraye' but now quite rare.
- *Hydrangea macrophylla* **'Mme Emile Mouillere'**, bred by Mouillere in 1909, previously given the AGM in 1992. Still one of the best whites, with large mopheads of serrate florets, continuously producing fresh flowers until the first autumn frost, that become touched with pink, green and palest blue as they fade. Better in half-shade, it layers and spreads creating a natural, tumbling display amid its companion plantings, rooting as it goes. My 40-year old plant, never pruned, has scrambled up 2m (6ft) into an old camellia and its white heads pour down from this height like a drift of snow to the lawn, the whole some 4m (12ft) across at the front of the bed. I have had to take a metre of growth off the front to prevent its consuming more and more lawn. An easy, generous and indispensable plant.
- *Hydrangea macrophylla* **'Lanarth White'**, a seedling from Lanarth, a Williams garden, also got its first AGM in 1992. A white lacecap that does well in full sun, flowering continuously over the summer. Howarth-Booth grew it on a south slope in sun, on greensand, where it grew to 0.5m (1½ft) and flowered continuously. It can reach 1.25m (4ft) in friendlier conditions. Flowers are small to medium size, bigger in some shade. A tough and easy plant, with pale-green foliage.
- *Hydrangea macrophylla* **'Tokyo Delight'** bred by Rothschild in 1930 from plants found in Japan, also got its first AGM in 1992. Another white lacecap with pure-white ray flowers that age pink, and surrounding pink fertile flowers in medium-size corymbs on an upright sturdy bush to 1.5m (5ft). Very free-flowering, it will take full sun or part-shade, and offers some autumn colour.
- *Hydrangea macrophylla* **'Veitchii'**, introduced in 1881, also with an AGM in 1992. One of the all-time greats: vigorous, reliably hardy and after 140 years still a first choice for many, including me. A great plant for beginners, very easy to grow, it has large, white lacecap flowers with three-quarter florets flanking blue fertile flowers in acid soil. It needs space as it spreads quickly, especially in part-shade with good soil, 1.5 × 3m (5 × 9ft). My 35-year-old plant is 5m (15ft) across.
- *Hydrangea macrophylla* **'Mariesii Perfecta'** is the 1904 Lemoine introduction, probably better known now by the Howarth-Booth name of **'Bluewave'**. It is a classic lacecap macrophylla with strong stiff stems on a stout bush that holds the flowers well above the foliage. At its best it is an attractive mid-blue in highly acid conditions; but it is a difficult 'bluer', and mostly seen as a medley of blue, lilac, pale violet and grey shades with a rather faded look. Though still commonly seen, especially by the sea, I think it is overpraised and overrated as a garden plant; there are better, more vivid choices – among the Tellers, for example.

Hydrangea macrophylla 'Générale Vicomtesse de Vibraye'.

Hydrangea macrophylla 'Mme Emile Mouillere'.

Hydrangea macrophylla 'Mariesii Perfecta' syn. 'Bluewave'.

Hydrangea macrophylla 'Altona'.

Hydrangea macrophylla 'Lanarth White'.

Hydrangea macrophylla 'Mariesii Lilacina' syn. 'Lilacina'.

Hydrangea macrophylla 'Bläuling'.

Hydrangea macrophylla 'Veitchii'.

Hydrangea macrophylla 'Blaumeise'.

- *Hydrangea macrophylla* **'Mariesii Lilacina'**, bred by Lemoine in 1904, with an AGM in 1992. One of the original Lemoine seedlings from Mariesii, and clearly a hybrid, probably with *H. serrata*. Its narrow foliage has long, pointed tips and is tinted purple on a 1.5m (5ft) bush that will tolerate poor conditions, including lack of soil moisture. Flowers are never fully blue, but settle for a purplish-mauve/pink, which varies depending on the acidity of the soil.
- *Hydrangea macrophylla* **'Bläuling'** bred at Wädenswil, one of the original Teller Series from 1984. Named in German for a blue butterfly, not a bird like most other Tellers, it well deserved its 1992 AGM. A lovely soft-blue to shell-pink lacecap with quite large ray flowers surrounding pale-blue fertiles. Blues easily. Quite compact, up to 1m (3ft) and produces a few late flowers.
- *Hydrangea macrophylla* **'Blaumeise'**, also one of the Tellers from Wädenswil, the German name for blue tit, launched in 1979 and given an AGM in 1992. The best known blue Teller with large, deep-blue bold ray flowers in an overlapping ring. Also labelled in some English garden centres as 'Teller Blue' or 'Blue Sky'. Vigorous, robust and hardy with strong shoots to about 2m (6ft) and a good constitution. Wädenswil rightly says *besonders empfehlenswert!* (specially recommended). A first-class plant.

A selection of other proven gardenworthy Teller cultivars

Although they did not receive an AGM at the latest trial, many other lacecap *H. macrophylla* cultivars in the Teller Series are top-choice garden macrophyllas in northern European conditions. All in this series were in practice the first group of free-flowering, reliably hardy garden lacecaps to be deliberately bred and made available.

- *Hydrangea macrophylla* **'Libelle'** (Wädenswil Teller, 1964) is a clean white lacecap of overlapping ray flowers, sometimes in a double row surrounding blue fertiles, with one or two fertiles scattered through the inflorescence, producing flowers steadily over a long season. Not as robust as other Tellers, and with more lax and floppy shoots, it does better in shade and with some shelter.
- *Hydrangea macrophylla* **'Nachtigall'** (Wädenswil Teller, 1979). The German for nightingale, this is the darkest royal blue to violet Teller, a strikingly rich colour in acid soil, and in acid conditions can be stunning. The intense colour needs time to settle, so be patient. In lime conditions it offers a dark pink. Vigorous, with strong shoots, it reaches 1.5m (5ft) and does best in shade and some shelter. Its colour and vigour warrants an AGM in my view.
- *Hydrangea macrophylla* **'Möwe'** (Wädenswil Teller, 1964) named seagull in English, this plant is synonymous with *H. macrophylla* 'Geoffrey Chadbund'. Red to reddish-purple ray flowers sometimes form a double row on a big corymb. It makes a big, open bush up to 1.5m (5ft) that holds its shape. Reliable.
- *Hydrangea macrophylla* **'Kardinal'** (Wädenswil Teller, 1987) appropriately named after the famously red 'Cardinal', this offers a crowded row of large, vivid-red ray flowers with slightly concave and pointed florets. Free-flowering over a long season with flower buds in most upper leaf axils. Foliage is slightly glossy and gives a healthy look to a compact bush of about 1.5m (5ft), which prefers some degree of shade.
- *Hydrangea macrophylla* **'Rotschwanz'** (Wädenswil Teller, 1987). The German name for the redstart, this Teller offers bright, dark-red ray flowers with a distinct dish shape, narrow and longish giving a star-shaped effect as they encircle white fertile flowers. Growth is sturdy, forming a compact shrub a little over 1m (3ft) that is very free-flowering from axillary buds, and has smaller leaves than most with a slight gloss. Wädenswil says *Liebhabersorte* – a variety for hydrangea enthusiasts. This is an easy, compact, high quality lacecap, excellent for pots as well as the garden and a personal favourite.
- *Hydrangea macrophylla* **'Bachsteltze'** (Wädenswil Teller, 1987), wagtail in English, is quite a compact 1m (3ft) high with a neat flower: a ring of white ray flowers around dark blue fertiles. It can flower from axillary buds, among dark green slightly glossy leaves.
- *Hydrangea macrophylla* **'Fasan'** (Wädenswil Teller, 1979), pheasant in translation, is a bright-red lacecap with usually more than one row of large ray flowers. Strong stiff shoots reach around 1.25m (4ft). Can be more difficult than most to root from cuttings.
- *Hydrangea macrophylla* **'Gimpel'** (Wädenswil Teller, 1987), the bullfinch of the series is rose-red but in acid soils a stunning violet-blue, in quite dramatic contrast with the white fertile flowers, on a strong, hardy, rounded bush of around 1.25m (4ft). It can also flower from lateral buds.
- *Hydrangea macrophylla* **'Rotdrossel'** (Wädensivil Teller, Wädenswil 1987). Named after the redwing, this is another bright dark red, with one or two rows of overlapping florets on a large lacecap flower. Leaves are toothed and dark green with a slight sheen. A strong grower, up to 2m (6ft).

Hydrangea macrophylla 'Nachtigall'.

Hydrangea macrophylla 'Kardinal'.

Hydrangea macrophylla 'Bachsteltze'.

Hydrangea macrophylla 'Fasan'.

Hydrangea macrophylla 'Gimpel'.

Hydrangea macrophylla 'Rotdrossel'.

Hydrangea macrophylla 'Möwe'.

Hydrangea macrophylla 'Rotschwanz'.

Hydrangea macrophylla 'Libelle'.

Remontancy and continuous flowering of *Hydrangea macrophylla*

Recent research has been done in the USA on the initiation and development of hydrangea flower buds, particularly in relation to the production of remontant (repeat flowering or reflowering) cultivars of *H. macrophylla*. Knowledge of the factors governing the ability to produce a second flush of flowers on shoots of the current year remains somewhat scant, but achieving some remontancy is seen as a major factor in the revival of interest in the USA in *H. macrophylla* as a garden plant, especially in the more marginal climatic zones. The pioneer remontant cultivar *H. macrophylla* 'Endless Summer' (found nameless in a garden in the USA), for example, during the first three years after its release, following an effective marketing and promotion campaign, sold record numbers of plants. This remontant dream plant was hailed as a breakthrough partly because the common understanding had been that *H. macrophylla* only flowers on shoots made the previous year, which means flower buds have to survive the winter for next year's flowers to appear. As a consequence, in harsh climates where low temperatures are the norm, or late spring or early autumn frosts common, *H. macrophylla* cultivars were held to be scarcely worth growing, as the tip buds and upper axillary buds are frequently killed off before they can flower. The remontant cultivars have the ability to flower on regrowth shoots of the current year, like a herbaceous plant. This reblooming capability was thought to be a unique genetic trait enabling *H. macrophylla* cultivars to be grown and flower satisfactorily in cold zones. The remontant *H. macrophylla* 'Endless Summer' was launched in 1998, millions were produced and new crosses made to raise other potential remontants. At the time of writing, *H. macrophylla* 'Blushing Bride', 'Rosy Summer' and 'Twist 'n Shout' have been introduced as putative harbingers of a whole new generation of reblooming hydrangeas.

As part of the research to check reflowering potential, flower buds were sampled at different times and from different locations on the plant, and the resulting data suggested that some cultivars may have little or no requirement for a specific daylight length or temperature to induce flowering. This is a very interesting conclusion that might relate to my own empirical observations about the impact of shade on freedom of flower. It has always been understood that a hydrangea planted in full exposure to sun, all other factors being equal, will produce a profusion of flowering shoots compared to a plant in deep shade.

This is usually a given among hydrangea growers: light induces flower production. However, I have been growing a dozen cultivars of *H. involucrata* for many years, and most show a remarkable capacity to flower freely in deep shade. I had sited many of these as young plants in a degree of shade because in the literature they were always described as of doubtful hardiness and top cover provides protection from both untimely frost and excessive summer heat. But since planting, after two decades their sheltering trees have grown and spread to the point where most of these now mature *H. involucratas* see no direct sunlight, with some in the intensely dark shade created by large magnolias. Most notably, my plants of *H. involucrata* 'Viridescens', shaded to the south and west by a very large *Magnolia sargentiana robusta* and tucked under small trees of *Rhododendron davidsonianum* and to the north by 12m (36ft) trees of *Acer palmatum*, scarcely receive a flicker of direct light and are now virtually growing in the dark. In spite of this, as I

Hydrangea involucrata 'Viridescens': flowering in the leaf axils of strong current year shoots.

Hydrangea involucrata 'Hortensis' is strongly remontant.

Hydrangea serrata 'Shirofuji' creates fresh flowers continuously from June to November, and careful examination shows each new emerging shoot bearing a flower bud; no single shoot can be found anywhere on this compact, delicate shrub that does not have a flower bud.

said in Chapter 5, every shoot carries a typically conspicuous, large, involucrate round flower bud, including on basal shoots of the current year deep within the plant. To flower from buds on shoots of the current year in this way in deep shade is a remarkable phenomenon – and plants can do so even when shoots are shorter than normal in quite dry conditions, with tree-root competition. This capability should warrant some attention from the hybridists currently exploiting remontancy in the *H. macrophylla* 'Endless Summer' group of 'everblooming' plants. Other cultivars in deep shade I have also seen flower freely in this way are *H. involucrata* 'Plena' and *H. involucrata* 'Handemari'; indeed all cultivars of *H. involucrata* appear to share this ability.

Both Bean and Haworth-Booth refer to the capacity of *H. involucrata* 'Hortensis' (for many years the sole representative of the species in the UK) to flower on shoots of the same year if the top growth was killed to the ground by frost; and they recommended it for a cold garden on the premise that you would always see some flower, even after the most aggressive winter. They were aware that these plants are truly remontant – as well as exceptional shade-bearers.

Another cultivar of a different species that has a remontant capacity and flowers continuously from June to November without drawing breath, is the old *H. serrata* cultivar 'Shirofuji'. This is a small plant, peeking out from a shady border in my case, but with a clear sky to the north. It is impossible to find a shoot without a flower bud. This is a treasure of a plant, much admired for its compact growth, delicacy and incredible freedom and continuity of flower.

Since *H. macrophylla* and *H. serrata* are closely related species (McClintock placed *H. serrata* as a subspecies of *H. macrophylla*), it should be possible to transfer the strong continuous flowering gene of little 'Shirofuji' to the massive flower potential of *H. macrophylla*. It is surprising that the 'reblooming' fanciers have not picked up *H. serrata* 'Shirofuji' as a possible parent, as it is likely to be perfectly compatible with *H. macrophylla* cultivars, but they seem fixated on macrophylla alone. There are already hybrids between *H. macrophylla* and *H. serrata*, including some first-class intermediate forms with macrophylla-sized flowers and improved hardiness.

I have also noticed a continuous flowering feature in some other seedlings of *H. serrata*. Some of my own hybrid seedlings at White House Farm flower all at once, making a great show with a 'wow' effect, the whole bush covered in flowers at a stroke, while others flower steadily and continuously, but less prolifically, deep into the autumn. In the latter instance, one corymb may be fading while an adjacent shoot is bearing a scarcely formed flower bud with possibly six weeks between them. These two habits of bloom belong to quite distinct types of *H. serrata*. The former, one-time copious flowering form may reflect some *H. macrophylla* contamination.

In fact, remontancy in hydrangeas is not a new phenomenon, nor a new discovery. There is an old generation of 'rebloomers'. Some 70 years ago, Michael Howarth-Booth in his 1950 book *The Hydrangeas* published a list of *H. macrophylla* cultivars he particularly recommended, giving what he selected as excellent garden varieties a single star, with double stars for a few: it turns out that most of those with double stars were essentially remontant macrophyllas, capable of flowering on new wood of the same year.

He specially selected and double-starred these macrophyllas because, as he put it:

> For hydrangeas grown as outdoor flowering shrubs, freedom of flower is the first essential. The double-starred varieties are so free-flowering that every branch usually carries several corymbs, *and new shoots will often flower in the same season, like a herbaceous plant. Thus if the top is killed by frost, flowers may still be had* provided that any excess numbers of new shoots are removed and frost damaged wood is cut back to healthy growth. [My italics]

What I have observed is that this remontancy in Howarth-Booth's double-starred plants may be highly dependent on optimal growing conditions. While all flower freely from axillary buds below the terminal bud, I have yet to observe genuine remontancy – flower buds forming and flowering on the same year's new growth produced from the base.

In my own garden, my ten-year-old plant of 'Endless Summer' has typical standard blue or pink mophead flowers, and is said to produce flowers on basal shoots of the current year if tip buds are killed. That I have yet to see this happen may be because it is never cut back by cold. In my garden it is an average bog-standard blue mophead macrophylla, and has not produced flowering shoots from the base. My guess is that to produce new shoots of sufficient vigour to carry flower buds

Hydrangea macrophylla 'Endless Summer', the original remontant variety in the USA.

Hydrangea macrophyllla 'Felina Blue' remontant in optimal greenhouse conditions at Signature Hydrangeas.

Haworth-Booth's double-starred *Hydrangea macrophylla* cultivars with remontant potential

The fifteen double-starred varieties Haworth-Booth listed (many described in this chapter):

Altona	Enziandom
Europa	Fischers Silberblau
Imperatrice Eugenie	Générale Vicomtesse de Vibraye
Mme Emile Mouillere	Hamburg
Heinrich Seidel	Kluis Suberba
Niedersachsen	Preziosa
Violetta	Vulcain
Westfalen	

the same year, growing conditions have to be favourable, especially with moisture at the root, and apart from not having flower buds winter-killed, my own plant is perhaps too dry and in too much sun to have enough new shoot muscle to throw up remontant 'herbaceous' flowers.

In this context, it is interesting to note that certain varieties of *H. macrophylla* that have *not* yet been particularly selected for remontancy, if grown in optimal conditions to produce plants of vigorous and healthy growth, will produce flowers from basal shoots of the same year after the main flowering has finished. I saw a plant of *H. macrophyllla* 'Felina Blue' growing in ideal conditions in a shade tunnel at Signature Hydrangeas in Kent, with automatic irrigation and generous quantities of Osmacote fertiliser, with the first flowers appearing in June already faded and a second flush of new flowers appearing on two fresh 'herbaceous' shoots from the base, looking absolute perfection in mid-August.

Some further cultivar selections

There are some 350 cultivars listed in the current issue of the RHS Plant Finder available from nurseries in the UK. There will be many hundreds more for those with the resources and will to trawl nursery catalogues and the internet, so any selection here will only be partially representative, a mere glimpse of the vast range available, drawn from my personal experience of growing them or observing them in other gardens as good garden plants. Howarth-Booth estimated that there were already some 500 macrophyllas available in the 1970s. Many of these will have fallen by the wayside and no longer be in cultivation; and there is equally no guarantee that all the modern varieties mentioned will manage to stay the course in the testing conditions of the open garden and public taste. Most of those listed below are in the RHS Plant Finder's available 350.

- *Hydrangea macrophylla* **'Alpengluen'** (Brugger, 1950) synonymous with 'Alpine Glow', a translation, this grows to a compact 1.25m (4ft) and is a medium to large bright rose red mophead with glossy foliage.
- *Hydrangea macrophylla* **'Zorro'** (van Zoest, 2001) is a branch sport from 'Blaumeise', so the flowers are identical, with pale-blue fertiles and broad, tepalled bright-blue ray flowers. The big difference is that the strong stout shoots are a conspicuous black or dark purplish brown on a robust bush.
- *Hydrangea macrophylla* **'Klaveren'** (Pierre Michel-Kerneur, 1990) is a striking lacecap raised in Brittany with rich pink or purple/lavender ray flowers almost round, with a deckle edge, serrations rounded. A robust bush up to 1.25m (4ft). Foliage dark green. First rate.
- *Hydrangea macrophylla* **'Blauer Prinz'** (Matthes, 1925) synonymous with 'Blue Prince', a translation, this small 1m (3ft) shrub has pointed florets of a vivid blue. One of the best blues and good in autumn too.
- *Hydrangea macrophylla* **'Dr Jean Varnier'** (Shamrock, 1994) is a sport from Lanarth White, with pale-pink or washed-lilac ray flowers on a medium-sized hardy bush and, like its parent, free-flowering and sun-tolerant. Excellent for a small garden.
- *Hydrangea macrophylla* **'Enziandom'** (Steiniger, 1950) synonymous with 'Gentian Dome', a translation; one of the best dark blues in acid conditions and a good size corymb. A strong compact shrub to 1.25m (4ft) with good autumn colour.
- About the same depth of blue is **'Mathilde Gutges'** (Steiniger, 1946) and perhaps preferable, as it is a more compact grower and quicker and easier to 'blue'. An excellent blue for the smaller garden and highly recommended.
- *Hydrangea macrophylla* **'Europa'** (Schadendorff, 1931) was an AGM in 1992. Pale-blue mophead with good-sized fimbriate flowers on a tall vigorous bush up to 2m (6ft) high and with a tendency to spread and gangle. Some autumn colour.
- *Hydrangea macrophylla* **'Fischers Silberblau'** (Fischer, 1930) synonymous with 'Fischers Silver Blue', a translation, this is a pale-blue, free-flowering dwarf macrophylla under 1m (3ft) that fades in sun, so is better with some shade. A double-starred Howarth-Booth variety that can flower on shoots of the same year, like a herbaceous plant, though I have not seen this happen.
- *Hydrangea macrophylla* **'Hamburg'** (Schadendorff, 1931) is a sister-seedling of 'Altona' AGM, and of similar make-up, with deep-blue, large mopheads turning red in the autumn. A large plant that grows to 2m (6ft), spreads and is best with some shade.
- *Hydrangea macrophylla* **'Love You Kiss'** (Yatabe, 1990) is a picotee lacecap with white florets edged with pink. A major feature is the conspicuous red young foliage on a medium-sized bush of about 1m (3ft). An attractive plant in spite of the name.
- *Hydrangea macrophylla* **'Mousmée'** (Cayeux, 1967) is a large, spreading lacecap with informally scattered and distinctive ray flowers that never turn true blue, even in a low pH. They open cupped and deep lilac/purple with a white eye, then mature to a flat mauve/pink, which in acid conditions is washed blue/purple with reddish streaks. The flat tepals overlap, creating an attractive, almost circular flower. Prefers some shade. An appealing persona different from the norm.
- *Hydrangea macrophylla* **'Quadricolor'** is named for its display of leaves in four-colour variegations – dark-green, sea-green, gold and white. This compact plant is about 1.25m (4ft) in height and has distinctive loose lacecap white ray flowers with separated florets that do not overlap. Often confused with 'Tricolor', 'Nigra' and 'Mandshurica', which lack the yellow variegation in the foliage; if you fancy black stems, these have probably been superseded by more recently introduced varieties such as 'Zorro'.
- *Hydrangea macrophylla* **'Sabrina'** (van der Spek, 2002) is a mophead in the 'Dutch Ladies' Series with picotee flowers, white with a pretty pink edging. A compact plant, best in some shade but enjoys good light to flower well.
- *Hydrangea macrophylla* **'Selina'** (van der Spek, 2002). Another from the 'Dutch Ladies' Series: compact, up to 1m, with strong upright shoots with red/bronze tints to the young foliage, and light red lacecap heads with scattered ray flowers. Deep rich-purple in acid conditions, but not a ready bluer. Happy in sun so long as there is moisture at the root.
- *Hydrangea macrophylla* **'Sandra'** (van der Spek, 2002) is also in the 'Dutch Ladies' Series. A plant around 1m (3ft) tall with five to six evenly-spaced picotee ray flowers, white with a pink edge, on shapely serrated florets round a mass of white fertile flowers. Delightfully delicate and best in some shade.

- *Hydrangea macrophylla* **'King George'** (H. J. Jones, 1927) is one of the few cultivars still in cultivation that were bred by the Chelsea Gold Medallist English breeder a hundred years ago. It is a typically sturdy shrub to 1.5m (5ft) with large serrate florets in a crowded head, either rose pink or a mid-blue. It received the RHS Award of Merit in 1927 as a vase for exhibition purposes.
- *Hydrangea macrophylla* **'Rosita'** (Draps, 1964) has crowded flowerheads packed with smaller florets of a reddish pink. Useful because it flowers early, with some continuity.
- *Hydrangea macrophylla* **'Soeur Therese'** (Gaigne, 1947) – the full name is 'Petite Soeur Therese de l'Enfant Jesus' though this is always abbreviated today. An excellent mophead up to about 1.2m (4ft) with pure white flowers. Prefers some shelter and shade to be at its best.
- *Hydrangea macrophylla* **'Renate Steiniger'** (Steiniger, 1964) is another early flowering mophead of relatively compact habit up to 1.2m (4ft). Large flowers with large florets and a relatively ready 'bluer', a rich blue to purple blue in acid soils, or dark pink in alkalinity. Van Gelderen says it may be the best blue hortensia.
- *Hydrangea macrophylla* **'Beauté Vendomoise'** (Mouillere, 1908) is another 100-year-old French winner, still prized today for its large florets, perhaps the largest of all, up to 8cm (3in) across and barely washed with pink or blue, draped informally round a loose corymb of fertile flowers, the whole up to 25cm (10in) across. The florets narrow sharply at the base, leaving a distinctive gap between each petal. It makes a large bush, especially in shade, exceeding 1.5m (5ft).
- *Hydrangea macrophylla* **'Izu No Hana'** (1970) is found in the wild on the Izu peninsula, south-east of Tokyo. A plant of great character with crowded large heads of bright-blue fertile flowers, ringed by six or seven starry, small, double ray flowers of a paler blue/purple, the whole thing in pink on an alkaline soil. It will take plenty of sun and make a compact bush around 1.25m (4ft).
- *Hydrangea macrophylla* **'Mikawa Chidori'** (Japan) looks like a branch sport from a typical *H. macrophylla* (*see* image on page 88) and is another Japanese plant of great character, exceptionally free-flowering even in shade and an easy bluer. The corymbs are packed with small, starry, pointed sepals and are long lasting. A plant of individual form and appeal.
- *Hydrangea macrophylla* **'Violetta'** (Haworth-Booth, 1950s) is a two-star H-B introduction that will go blue in good acid conditions after a long wait, but usually remains a violet purple, with a mix of colours in this zone. It is, usefully, early to flower with weather-resistant flowers and excellent continuity, so the mix of maturing fading flowers is enlivened by a few fresh late summer blooms.
- *Hydrangea macrophylla* **'Harry's Red'** (Gyselinck, 1960). Sometimes known as Harry's red topper, this is a small shrub to about 0.5m (1½ft), with rich-red, smallish mophead corymbs freely borne. In acid conditions it is a reluctant bluer, with a wide colour mix of red, purple, violet flowers. It is comfortable here in Kent in full sun, with long-lasting flowers. A very good front-of-border plant.
- *Hydrangea macrophylla* **'Hanabi'** (Nakamura, 1960) is also listed as 'Fireworks', a translation. It's true that it does look something like a bursting firework, with starry white lacecap flowers with pointed sepals on very long pedicels looking upwards, as if exploding from the rim of the generous platform of fertile white flowers. It makes a significant free-growing, open shrub of about 2m (6ft).
- *Hydrangea macrophylla* **'Albrechtsburg'** (Nieschutz, 1983) is a rarely seen, first-class pink mophead, with conspicuously frilled flowers. It is a clean pink and slow to blue, opening from cream juvenile flowers. A strong shrub of 1m plus (3ft+) and one of the very best pinks.
- *Hydrangea macrophylla* **'Hobella'** (Hofsteder, 1994) is one of the Hovaria (Kaleidoscope) Series and another excellent pink lacecap hydrangea. Quite compact – I have it in a large pot for which it is ideal. The delicate pink flowers age in the autumn to a greenish tone mottled and edged red, so it stays attractive over a long season.
- *Hydrangea macrophylla* **'Cote d'Azur'** (Rampp, 1986) is part of the Cityline Series. A dramatic dark blue in acid conditions, but slow to 'blue' and usually seen as a deep violet/purple. Freely borne, smallish mopheads on a compact plant to 1.25m (4ft). A striking colour at its best.
- *Hydrangea macrophylla* **'Marechal Foch'** (Mouillere, 1924) is an old blue favourite and one of the best deep-gentian blues. Not a vigorous plant and the flowers enjoy some shade, which inhibits fading. Starts early, with good continuity on a 1m plant. Good in a pot.
- *Hydrangea macrophylla* **'Kay Leslie'** (Bullivant, 1990s) may not yet be available commercially, but is too good to omit from this list. A deep rose-red, flat corymb that does not fade even in full sun, where it is exceptionally free-flowering over a long season. An iron-clad cultivar for the open garden.
- *Hydrangea macrophylla* **'Princess Diana'** (Garden Express, 2017) synonymous with 'Wow Time' from the You and Me Series. A new mophead with a very different appearance, the big flowerheads composed of double, star-shaped florets with crowded narrow pointed petals and with good continuity. It is a typical compact macrophylla of great individuality.
- *Hydrangea macrophylla* **'Fragola'** (Doll, 2012) Dolce Series. A loose, flat mophead with lacecap tendencies. A strong strawberry-red with large florets on a compact bush. A bright colour as a red, but transforms into a good purple/blue in acid conditions.
- *Hydrangea macrophylla* **'Mariesii Grandiflora'** is the white version from the original Lemoine raising and also known by the Howarth-Booth name of 'Whitewave'. It is an excellent white lacecap with squared sepals that are

Hydrangea macrophylla 'Alpengluen'.

Hydrangea macrophylla 'Zorro'.

Hydrangea macrophylla 'Klaveren'.

Hydrangea macrophylla 'Dr Jean Varnier'.

Hydrangea macrophylla 'Enziandom'.

Hydrangea macrophylla 'Mathilde Gutges'.

Hydrangea macrophylla 'Fischers Silberblau' syn. 'Fischers Silver Blue'.

Hydrangea macrophylla 'Hamburg'.

Hydrangea macrophylla 'Quadricolor'.

Hydrangea macrophylla 'Selina'.

Hydrangea macrophylla 'Sandra'.

Hydrangea macrophylla 'Sabrina'.

Hydrangea macrophylla 'Love You Kiss'.

Hydrangea macrophylla 'Mousmée'.

Hydrangea macrophylla 'King George'.

Hydrangea macrophylla 'Soeur Thérèse'.

Hydrangea macrophylla 'Rosita'.

Hydrangea macrophylla 'Beauté Vendomoise'.

Hydrangea macrophylla 'Hanabi'.

Hydrangea macrophylla 'Izu No Hana'.

Hydrangea macrophylla 'Mikawa Chidori'.

Hydrangea macrophylla 'Violetta'.

Hydrangea macrophylla 'Kay Leslie'.

Hydrangea macrophylla 'Côte d'Azur'.

Hydrangea macrophylla 'Harry's Red'.

Hydrangea macrophylla 'Marechal Foch'.

Hydrangea macrophylla 'Fragola'.

Hydrangea macrophylla 'Princess Diana'.

Hydrangea macrophylla 'Mariesii Grandiflora'.

Hydrangea macrophylla 'Hobella'.

Hydrangea macrophylla 'Westfalen'.

Hydrangea macrophylla 'Le Cygne'.

Hydrangea macrophylla 'Sea Foam'.

Hydrangea macrophylla 'Albrechtsburg'.

Hydrangea macrophylla 'Merveille Sanguine'.

shaded with pale pink as they age. It is a vigorous 1.5m (5ft) plant that is not as often seen in gardens today as it merits; I think *H. macrophylla* 'Vietchii' is generally more available. Sometimes the two are confused in the trade.

- *Hydrangea macrophylla* **'Westfalen'** (Wintergalen, 1940) is a Howarth-Booth favourite and one of his two-star selections. It is an excellent red mophead, compact in habit and according to H-B fully remontant and thus 'flowers often on shoots of the year in the same season *if well cared for*' [my italics].
- *Hydrangea macrophylla* **'Merveille Sanguine'** (Cayeux, 1939) translates as the bloody marvel, in both senses, as it is a difficult plant and very hard to please. Like Longfellow's little girl with a curl – 'when she was good she was very good indeed, but when she was bad she was horrid'. A sport from the pink Merveille, this is the darkest red yet raised, with dark red-purple young foliage, but the flowers rot if it is too wet and curl if it is too dry; too much shade and flowering is inhibited, too much sun and growth is inhibited. In acid conditions a rich purple/red. Always worth trying as the colour is striking.
- *Hydrangea macrophylla* **'Le Cygne'** (Cayeux, 1919) or the swan – an old white mophead with great vigour, to 1.5m (4ft). Large florets in a tousled head and, with age, the white becomes stained and mottled with deep pink. A reliable and easy white cultivar.
- *Hydrangea macrophylla* **'Sea Foam'** (Haworth-Booth, 1950s). Michael Haworth-Booth found this as a reverting branch sport of 'Joseph Banks' on the Isle of Wight, concluding that it represented the wild coastal species of Japan and naming it as a form of what he dubbed *H. maritima*, now regarded as a *nomen nudum* and a form of *H. macrophylla*. It is also reasonable to conclude that the mophead 'Joseph Banks' is itself an anomalous branch sport of this wild form found by Banks in China, but a Japanese import. While it thrives at the seaside, it is less effective inland, with relatively small ray flowers, sparsely borne, doubtfully hardy and of little garden value. Its interest lies in its history.

Hydrangea serrata – sophisticated summer chic

Hydrangea serrata is much less widely planted in Europe than *H. macrophylla*. It is much less well known, although its popularity is increasing steadily year by year as it becomes better appreciated and new cultivars are introduced. The number of cultivars available in the RHS Plant Finder, for example, has more than doubled in the last ten years, with now some 90 to choose from. In a 2013 review, seven *serrata* cultivars were awarded the AGM.

Compared to the hortensia, *H. serrata* could not be more different in character. It is much smaller in stature, rarely seen above about a metre in height, and an altogether more dainty and delicate-looking hydrangea: a twiggy, slender-branched shrub with smaller, narrower and finely- or coarsely-toothed leaves and smaller flowers. It might be said to be more representative of Japanese taste: more nuanced, refined, elegant, understated. Most of the cultivars available are of Japanese origin, and with Japanese names, of which there has been an inflow of new arrivals in Europe over recent years.

Hydrangea serrata distribution

Although it looks delicate, *H. serrata* is in fact a tough, reliably hardy plant in all parts of the UK and the species most suited to our fickle climate. It is a mainly montane species, with a shorter growing season than the soft maritime hortensia, which makes it less vulnerable in cultivation to late spring and early autumn frosts. In the wild it is widespread in Japan, ranging from Hokkaido in the north right down to the woods and hills of Kyushu and Sikokyu in the south, with some presence on the island of Ullung-Do off the coast of South Korea. It also has a wide altitude range, from 1,500m (4,800ft) in mountain forest and on open ridges down to virtually sea level.

Hydrangea serrata in Takahata Fudo Temple, Tokyo.

Indigenous *H. serrata* growing naturally in a temple precinct.

The potential of *Hydrangea serrata* as a garden plant

This wide variety of habitat in the wild may be why *H. serrata* is such a variable plant, in flower, habit, foliage and constitution. Howarth-Booth broke out three separate species from *H. serrata* originating in the inland woodland and montane areas of Japan, namely: *H. japonica*, *H. acuminata* and *H. thunbergii*. However, as McClintock pointed out, the characters he identified were not based on morphology, but for the most part on ecological and cultural factors, such as shade-tolerance, flowering time, texture of leaves and habit – not characters on which a species' definition can be based. Many variations arise from natural sports or mutations, giving double flowers, sterile forms and a wide variety of colours, which accounts for the popularity of *H. serrata* in Japan, where it is deeply embedded in Japanese culture. In addition, many seedlings exhibit characteristics suggesting a hybrid origin with other associated species in nature, such as *H. macrophylla*, *H. scandens* and *H. hirta*. Some of these hybrid seedlings in the wild frequently have worthwhile garden merit.

A putative hybrid found in the wild : a cross between *H. serrata* and *H. hirta*.

Typical form of *H. serrata* in the wild, with a ring of pale-blue ray flowers hovering round an open disc of fertile flowers.

Japanese forms of *H. serrata* in a vase showing variations in flower form and colour, plus a dark-blue White House Farm *serrata* seedling.

In cultivation, crosses with *H. macrophylla* are also in evidence, revealing some compatibility and a close relationship. Notwithstanding this, and the McClintock treatment of serratas as a subspecies *of H. macrophylla*, I prefer to treat *H. serrata* as a separate species (albeit in the same Macrophyllae subsection) as their morphological and cultural differences are immediately evident.

Flower colour in the wide range of *H. serrata* cultivar selections ranges from white, through pale and rich pink, to rose red, purple and deep red, with further variety in shades of blue in acid soil conditions. Flowers are typically of the lacecap variety with a flat disc of small fertile flowers ringed by conspicuous ray flowers, like a hover of butterflies.

There are cultivars with single ray flowers in a variety of shapes, sometimes scattered randomly through the

Small compact form and shade tolerance make *H. serrata* ideal for border and underplanting.

inflorescence. There are doubles, some toothed, some entire and some with a mass of very narrow tepals to give a starry effect; and a few mophead and partial mophead types, small and neat. Selection becomes a matter of taste and inclination. *Hydrangea serrata* foliage is shapely and refined, the leaves relatively narrow and toothed, more than twice as long as wide, frequently with reddish or purple tints and often colouring to red and purple in the autumn. Most will flower from June onwards in the UK.

Most cultivars are small in stature, usually around 1m (3ft) or less, but there are a few cultivars up to 2m (6ft). Constitution is generally tough and cold-tolerant, but some cultivars, sometimes the most desirable, are fastidious and require attention to siting and soil to get the best results in the garden.

A combination of delicacy and toughness

The garden value of *H. serrata* comes from its wide range of flower types and shapes, from small mopheads to double-

The delicate-looking *H. serrata* 'Bluebird' is indifferent to the freezing winters of Nova Scotia.

flowered forms, to lacecaps; from its extensive range of colours, often changing dramatically over a long season, and from its hardiness and resistance to winter cold: in short it offers more versatile uses in the garden than *H. macrophylla*. It is also superb as a patio plant, viewed at close quarters, as it is more understated than the assertive *H. macrophylla*, more Mozart than Wagner, more flute than trombone.

My descriptions of the various and varied cultivars will focus on their garden value and support my belief that they are better suited to UK and even inland European conditions than macrophylla. In particular, although it looks delicate, it is in fact a tough, reliably hardy plant. It is mostly a montane species, so has a shorter, later growing season than the soft maritime hortensia, wood mostly ripening for the next year by August, which makes it less vulnerable in cultivation to early autumn cold snaps and the onset of winter. On a visit to

Autumn can be a colourful season for *H. serrata*.

Nova Scotia, where frequently *H. macrophylla* is cut to the ground and ungrowable, I learned to my astonishment that *H. serrata* cultivars thrive and flower well, coming unharmed through winter cold that falls into the minus twenties centigrade.

There is plenty of evidence that the appreciation and popularity of *H. serrata* is growing steadily year by year as its qualities are better understood and new cultivars are being introduced from breeding, naturally occuring hybrids and seedling forms in gardens, anc from the wild.

Definition of Yae and Temari

'Temari' is the Japanese term for a mophead, though this crude work-a-day European label does not properly reflect the more refined sense of its Japanese tradition.

It actually denotes a small fabric ball that young girls would play with before rubber or plastic balls were available. These were exquisitely hand-embroidered with colourful threads in infinite patterns, a beautiful example of traditional handicrafts, and are now mostly used for decoration, especially for New Year celebrations, like our own Christmas baubles. *See Hydrangea* 'Besshi Temari' later in this chapter for an excellent example of this in a hydrangea flower. The term 'Handemari' (*see H. involucrata* 'Handemari' in Chapter 5) is a looser form of Temari, with more fertile flowers – a kind of halfway house, with more of a mixture of ray flowers and fertile flowers providing a marginally less formal effect.

Hydrangea serrata – 'Temari' style of flower.

'Yae' is the Japanese term for 'double'. I understand it is written in Japanese as 'eight layers' and actually can signify an unspecified number of petal layers. 'Komochi 'seems to be an extension of the 'Yae' term used for double flowers, where a much smaller separate ring of florets is not tiered directly on top of the others and nestle in the arms of the outer, senior floret. It means in effect 'mother holding baby'.

A 'Yae' form, *H. involucrata* 'Yohraku', showing how ray flowers are produced continuously on top of each other over the summer like a daisy chain, turning from mauve to pale green as each matures to as much as an 'eight layers' flower build-up, each ending in a fresh mauve-white flower (right).

A 'Komochi'-style flower with baby florets nestling inside the main large floret.

Most cultivars are below or about 1m (3ft), but there are a few up to 2m (6ft). Whether massed at the front of a border or the edge of a copse, or as informal hedge, or lighting up the base of a wall or flanking a porch, or as a light and frothy foil to heavy evergreens, serratas provide colour, chic and charm for weeks on end. All are good for cutting for vase decoration and long-lasting if stems are kept short and vases kept in a cool place.

Growing *Hydrangea serrata* in gardens

Chapter 10 offers practical tips about growing hydrangeas in general, but there are issues specific to *H. serrata* that can influence its garden performance. Serratas have a tough and easygoing constitution, tolerant of most soils. Although they favour an open, fertile, well-drained, neutral or acid soil, they will grow perfectly well in a well-structured alkaline soil – though, as with all hydrangeas, excessively chalky soils may induce chlorosis. Good drainage is essential, and moisture at the root desirable through the growing season, for which a suitable mulch is beneficial. An acid soil, below pH6.5, is necessary for natural blue flowers, which are a function of available aluminium in the soil. (For more, *see* 'Soils' in Chapter 10). Blue can be achieved on neutral soils through treatment with aluminium sulphate or proprietary 'blueing agent'. Take care, as overdosing can cause significant, sometimes fatal, damage. It is good practice to 'go with the flow' and on alkaline soils simply select the best pinks, reds and whites.

Siting is important, though plants are perfectly winter-hardy. Continuous exposure to summer sun and high temperatures can cause flowers to wilt; and constant exposure to high summer temperatures in the thirties centigrade can burn and shrivel foliage. Serratas flower at the summer solstice when the sun is at its hottest and appreciate some shade protection. It is the leaf surface temperature that counts, to which irrigation at the root makes no difference.

Apart from the 'Beni' group, which open white and mature to red, there are some other excellent reds. This is a *serrata* hybrid, 'Garden House Glory': very free-flowering and effective in the landscape.

For shade, woodland is effective, but local partial shade in a small garden from, say, a large *Philadelphus* or other small tree is perfectly adequate. Such companions can also offer shelter from cold winds and some frost protection, while taking care to avoid too much root competition sucking moisture from the soil and denying it to the hydrangeas.

A first-rate White House Farm red seedling with little inclination to change colour in acid conditions.

A White House Farm seedling that flowers for four months.

H. serrata 'Tiara'.

Planting on, or at the top of, a sloping site helps to mitigate the effects of late-spring frosts. Even in a small garden, slight undulations or slopes can create a pool of denser, heavier cold air – the notorious frost pocket (*see* Chapter 10). But plants can be easily re-sited. With a compact root system they can be moved at any age with a good rootball, and if watered-in well and shaded temporarily, they can be moved at almost any time of year.

Many *serratas* offer stunning autumn colour as late as November.

Pruning requirement is minimal, but the oldest wood can be removed from the base, which thins out old, spent twigs and encourages new growth. If a plant becomes congested, it can be cut hard back, taking care to thin out the ensuing crowd of new shoots to allow wood ripening. I have taken a hedge cutter to several old, tired plants to excellent effect, with even one or two flowers that same season on the vigorous new growth.

Some recommended cultivar selections

White cultivars

Excellent white forms include **'Shirofuji'** AGM – the white hydrangea from Mt Ashitaka, an old volcano south-east of Mt Fuji – which grows to only 0.5m (1½ft), and produces ray flowers with a double row of creamy white florets continuously from June to frost. Few shoots are without flower. It is an immensely charming plant, which I have peeping out from below

Hydrangea serrata 'Shirofuji' AGM.

Hydrangea serrata 'Fuji No Taki', the 'Waterfall of Mt Fuji'.

Hydrangea serrata 'Hakucho' (swan) has three to five double ray flowers.

Hydrangea serrata 'Shirotae'.

Hydrangea serrata 'Shinonome'.

A lacy White House Farm *serrata* seedling.

'Hakusen'.

others at the front of a shaded border. It is excessively free-flowering. I sometimes play a game with visitors – if they can find a shoot without a flower or a flower bud they can take it as a cutting. To date, in spite of close inspection, no cuttings have ever been taken. It is the most continuously flowering hydrangea of any cultivar that I grow. Christopher Lloyd wrote after seeing it at White House Farm: 'However anti-hydrangea you may think you are, you couldn't fail to be endeared by one like this.' I would not be without it.

'Fuji No Taki' – the 'Waterfall of Mt Fuji' will grow to 1m (3ft) with more formal, individual double blooms congested with petals, each between 1 and 1.5cm across in multi-flowered heads. The Japanese know them as 'han-demari', a halfway house between lacecap and mophead. They are lime green in bud, maturing to cream, then white and fading again to pale green. It exudes charm for months starting in early June and for me is indispensable.

'Hakucho' has three to five double ray flowers, with white fertile flowers at their centre. It needs shade and is not as robust or free-flowering as others but, once established, has its own personality. Another hydrangea found on Mt Fuji by the Gotemba Noen nursery. Translates as 'swan'.

'Shirotae' is a term used poetically to refer to white – clouds, snow, cranes and flowers, for example. I have found the plant rather hard to please initially, slow to start but repaying care with a mass of starry, double, white flowers freely scattered across a 0.6m (2ft) bush over a long season. In good light it is a mantle of lace. It too benefits from shade, though with full light overhead it flowers more freely. It is similar to 'Shirofuji' but less easygoing for the gardener, as well as less continuously flowering.

'Shinonome' has conspicuously starry, single flowers just washed with pale blue, scattered over a domed inflorescence and distributed freely across a neat bush about 0.75m (2½ft) high and perhaps 1m (3ft) across. It is distinct, with individual character. The name means 'dawn clouds in the east'.

One of my as yet unregistered seedlings from an open-pollinated 'Maiko' is obviously a hybrid with *H. scandens*. It sends up shoots of about 0.5m (1½ft) each year and they flower from all the leaf axils down the stem, the slender branches wreathed in white flowers, slowly turning lime green into autumn. It is graceful and elegant in habit, the slender branchlets bowed gently under the weight of the flowers.

'Hakusen' is rather spotty in flower, and the ratio of flower to foliage is poor, even in good light, but the small, tight, congested buns of temari flowers, about 6cm (2½in) across, are distinctive. The individual florets are overlapping and larger than the norm in these plants. Perhaps it should have full sun to induce greater freedom of flower. It grows up to 0.75m (2½ft). 'Yae Hakusen' is a double form, again with attractive corymbs but rather spotty flowering.

Hydrangea serrata 'Fuji No Shirayuki'.

Hydrangea serrata 'Janis Joplin'.

Hydrangea serrata 'Sugimoto', free-flowering and very early.

Hydrangea serrata 'Amagi Amacha'.

'Fuji No Shirayuki' wanders gently about among evergreen Japanese azaleas and they make friendly and amiable companions, perhaps recognising that they are compatriots, each complementing and complimenting the other. It elbows its way around, keeping its head up to about 0.5m (1½ft), studded with starry, dainty white flowers over a long period. Another form from Mt Fuji.

'Janis Joplin' is a dense twiggy mound up to 0.75m (2½ft) covered in small, white lacecaps, with both ray and fertile flowers white. It was found in Japan by Corinne Mallet without a name and she named it for her favourite singer. It is generous with its flower when in full song.

'Sugimoto' flowers much earlier than most and makes a tight, dense bush of about 1 × 1m (3 × 3ft). It may be a hybrid with either *H. scandens* or *H. luteovenosa*. It has shapely, almost round, ray flowers and is exceptionally free-flowering. This, with its very desirable early flowering, makes it distinctive.

'Amagi Amacha' is a delightful small plant with myriad small, white lacecap flowers covering the bush, all light and air, a delicate spread of lace.

These small white serratas will not shout colour in the garden landscape like the *H. macrophylla*, but they sing their charm throughout the summer. At close quarters, along the edge of a path or part-shaded border they always stop delighted visitors who invariably ask where they can acquire them.

The 'Beni' (red) group

A second charming and easily recognisable group – we could call it the Beni group – are those that start with fresh white flowers, then flush and stain pink, finally turning red as they age, some colour-charting eventually at pure crimson. 'Beni' is a pigment used to dye fabrics and in cosmetics – a bright red, as used in Kabuki stage make-up.

The Beni serratas make bigger bushes than the whites, mostly exceeding 1 × 1.5m (3 × 5ft) across, and have the capacity to make a colour statement in the landscape, appropriately sited. The flower corymbs usually have some four to eight ray flowers.

The best known and most widely seen is **'Beni Yama'** AGM, which makes a vigorous, rounded shrub to about 1m (3ft) with some flower continuity, which means it displays individual flowerheads at various stages of development, mixing white with flushed pink, deep pink and red all on the same bush. This is an easy and accommodating plant and will tolerate a relatively open and sunny position, though may be best with some afternoon shade. The fertile flowers are lilac/mauve. It translates as 'red mountain'.

'Beni Yama' is sometimes sold in the trade as **'Beni Gaku'**, which is quite different in most respects, including lacking a willingness to grow freely. It can be a recalcitrant little beast though you try to forgive it as it is distinctively attractive. It is an old Japanese variety of unknown origin that has never been found in the wild, and is said to be the first of this type to be brought into Europe. The name means 'red frame'. The most obvious difference with 'Beni Yama' is that the fertile flowers are white, while those of 'Beni Yama' are blue/mauve; another is that the sterile ray flowers are generally larger, often with only

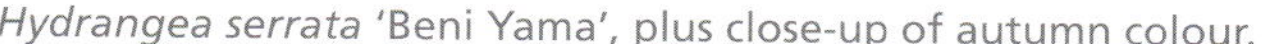

Hydrangea serrata 'Beni Yama', plus close-up of autumn colour.

Hydrangea serrata 'Beni Gaku': note the white fertile flowers.

Hydrangea serrata 'Kurenai' transitioning from white to red and fully mature in late summer.

three petals, the lower one larger than the others. It needs shade and moisture, though not too much of either.

Probably the finest red of these Beni-types, though again sometimes mixed up in the trade, is **'Kurenai'**. I find this also a weak grower, but have seen it do better in light, sandy soil. It is worth all the attention you can give it, as the final red colour of the delicate lacecap flowers in late summer is stunning, a most intense red, especially in good light and not too much shade. The ray flowers have either three or four sepals. It was found on the Abe Pass at 1,400m (4,480ft) in Nagano Prefecture, Japan.

'Aka Beni' is the most vigorous of this group, in light shade growing up to 2m (6ft). I also grow it in full sun among gallica roses where it appears happy, making a dense, upright bush of a little over 1.25m (4ft) and flowering more freely than in shade; and the flowers do not easily sunburn. It goes through the usual colour change process, starting white and ending up a deep pink.

'Beni Temari', a 'red ball', is the mophead version of this group, cultivated during the Edo period 1603–1868 but lost and rediscovered by Mr Yamamoto, President of the Nippon Hydrangea Association. It makes a mounded plant about one metre each way. The individual temari flowers are in a rather loose, more informal mophead moving from white to a soft pink flush at maturity.

The outstanding **'Grayswood'** AGM is probably a hybrid and not typical, as it will grow to a bit under 2m (6ft), with a rather loose, open habit. It was found in the UK in its namesake garden near Haslemere in Surrey. It is sometimes slow to start, but has a robust constitution, and will be happy in sun. It is one of the showiest garden cultivars. In full season, the plant is a medley of white, pink and red ray flowers at different stages of transition, with blue fertile flowers in an acid soil. The ray florets are pointed, shapely and distinctly toothed.

'Preziosa' AGM grows to around 1.5m (4ft), a dense twiggy bush, possibly a hybrid with *H. macrophylla* as it is

Hydrangea serrata 'Aka Beni'.

Hydrangea serrata 'Beni Temari' – the mophead version of this group.

Hydrangea serrata 'Grayswood' AGM.

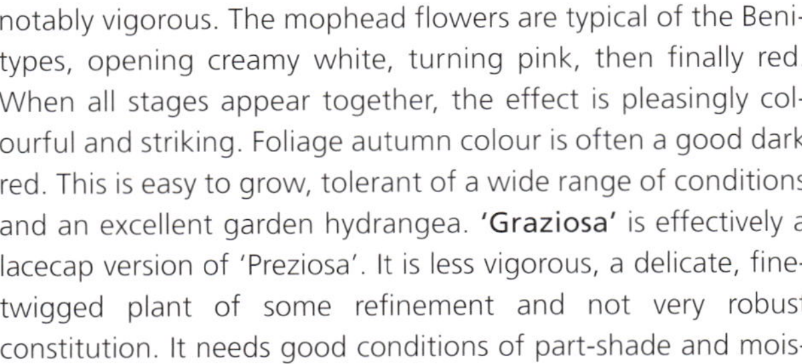

notably vigorous. The mophead flowers are typical of the Beni-types, opening creamy white, turning pink, then finally red. When all stages appear together, the effect is pleasingly colourful and striking. Foliage autumn colour is often a good dark red. This is easy to grow, tolerant of a wide range of conditions and an excellent garden hydrangea. **'Graziosa'** is effectively a lacecap version of 'Preziosa'. It is less vigorous, a delicate, fine-twigged plant of some refinement and not very robust constitution. It needs good conditions of part-shade and moisture to be at its best. The foliage is a dark green, with crimson autumn colour.

'Rosalba' AGM is a very old Japanese clone with a confusing history and a good number of synonyms. It is an easy and reliable plant with five to seven ray flowers, which turn from white with, eventually, some red staining and ultimately a bright crimson. The foliage is distinctive, a pale bloomy green on an upright bush to about 1.25m (4ft). The similar **'Juno'** is a more modern version and has more vigour, with a stronger constitution.

Hydrangea serrata 'Preziosa' AGM.

H. serrata 'Juno', similar to 'Rosalba'.

H. serrata 'Bluebird' AGM in Nova Scotia.

Some attractive blue forms

Some of the best simple blue lacecaps are of UK seedling origin, notably those raised from Japanese seed by Michael Howarth-Booth. These are blue in acid and pink in alkaline conditions, which is why it can be slightly misleading to name a hydrangea for its colour, as it is so variable in differing soil conditions. Some plants are slower to absorb aluminium and thus more reluctant than others to produce blue flowers, even when acid conditions are present.

'Bluebird' AGM is widely grown, with a neat, slightly domed inflorescense and mid-blue fertile flowers ringed by pale-blue or pink ray flowers, with rounded, entire, separated florets. It is an easy, tolerant plant, tough in low temperatures, so especially good for cold gardens. It comfortably tolerates –20°C. As it is often pink, it is an example of the dangers of naming hydrangeas according to colour. Can you imagine pink birds over the white cliffs of Dover?

'Diadem' AGM is similar, but a smaller plant and perhaps less free-flowering, with a paler flower. More than one clone circulates under this name in the trade. For me the award is borderline.

'Miranda' AGM is a gem and my touchstone for quality colour in blue *H. serrata*s. It is butterfly blue, and the only truly self blue, that is with both fertile and ray flowers more or less the same colour. They are nicely held above the foliage on a free-flowering shrub about 1m (3ft) high and as much across. It enjoys some sun and needs plentiful light to be at its best.

An unnamed sister-seedling that I was given by Michael Haworth-Booth many years ago from the same group of seedlings, I later introduced and named **'Tiara'** AGM. It is a more robust hardy plant, growing to 1.5m (5ft). In sun, the foliage is tinted purple and colours well in autumn. Ray flowers are soft blue or pink, very freely produced and effective over a long period. I gave it to a Brittany nurseryman, he promoted it well with large posters and found it hard to satisfy demand, and in France it was the most sought after of all his cultivars.

'Diadem' AGM.

'Miranda' AGM.

'Tiara' AGM in close up and wide shot.

Double-flowered forms

Apart from the double whites described above, there is a very good selection of double coloured cultivars. One of the best is **'Miyama Yae Murasaki'** AGM, which will achieve 1.5m (5ft) as a spreading bush. When the conspicuous blue/purple double ray flowers on long pedicels fade to purple, smaller flowers continue to open at the centre of each floret and scatter the whole inflorescence with fresh blue. In the USA, it is known as **'Purple Tiers'**. The Japanese name translates as double purple, and it originates in an area of high snowfall on the Horikoshi Pass in Kyoto.

'Mikata Yae' comes from Mt Hachibuse in Mikata county, Hiyogo Prefecture and has a very starry, double, dark-pink flower that is reluctant to turn blue but will ultimately do so under acid conditions. It is under 1m (3ft). Individual florets are large, about 3cm (1in) across, long and pointed, often freely borne and overlapping in the inflorescence to create a striking star-like effect.

'Kocho No Mai' originates from ancient woodland in the same area as 'Mikata Yae'. The name translates as the butterfly dance, with small, pale mauve/pink ray flowers nestled at the centre of a larger one, creating a pretty double effect in the 'Komochi' style. The ray flowers are narrow and pointed, and there are a few fertile flowers in an open, loose corymb, evoking the gentle movement of small butterflies. 'Kocho' is the literary term for butterfly, and the hydrangea probably refers to an ancient court dance first performed in 908 and said to have been put on for an emperor attending a boys' sumo match. This is today performed as a ritual traditional dance at Shinto shrines, usually by four boys dressed in white with green outer garments with a tiara headpiece and light butterfly wings of pasteboard. *Kocho no mai* is also a kind of magical performance dating from the 1700s using a folding fan to simulate fluttering butterflies

'Miyama Yae Murasaki' AGM.

'Mikata Yae'.

'Mikata Yae' in acid soil in Nova Scotia.

'Kocho No Mai'.

'Shishidanka'.

and then through sleight of hand, to turn them into 'whirling confetti'.

'Kisawa No Hikare' is perhaps the most eye-catching cultivar of all and would easily warrant specimen treatment with three or five plants grown together in part-shade to form a significant mass. The mature foliage is elegant, with long, pointed leaves, which are very dark, almost blackish green; the shoots are dark and the spectacular red/pink ray flowers flaunt stringy petals like huge spiders, ringing yellowish-green fertile flowers. It makes a small spreading bush of about 0.75m (2½ft), the whole a brilliant combination of flower and foliage. Indispensable.

'Aya Ezo Yae' is a delightful double-flowered cultivar, the flowers a combination of pale pink and a lovely soft, clear shell-pink, held on long pedicels in an open airy corymb with few, if any, fertile flowers. These pinks fade to a pale green, giving a nice combination of fresh and faded flowers over a long season on a sturdy rather upright 1m (3ft) shrub. An outstanding cultivar.

'Shishidanka' is a distinctive cultivar described by Phillip von Siebold in his *Flora Japonica* (*see* Chapter 1) but lost until rediscovered in the hills of Kobe in 1959. Now popular and widely available.

'Yae No Amacha' is a double-flowered blue or pink lacecap with a scattering of additional florets across an open corymb. An easy, long-flowering plant that will take full sun better than most, if moisture is available. It makes a compact plant up to 1m (3ft), is dense and twiggy, and makes effective ground cover with maturity. Amacha refers to its use to make a sweet tea.

'Kisawa No Hikare'.

'Aya Ezo Yae'.

So-called 'Amacha' hydrangeas for tea-making

Hydrangeas with 'amacha' in their name are used to make a special sweet tea. Fresh hydrangea leaves are washed, steamed, then dried to produce dark-brown tea leaves. This may be used for tea ceremonies, and there is a tradition to pour this tea over statues of the Buddha on the occasion of his birthday.

Care needs to be taken in sampling homemade hydrangea tea without good knowledge of the type to select, as some hydrangea leaves are toxic, containing cyanide. Such leaves, notably of *H. macrophylla*, may not be deadly but can cause severe stomach upset.

Some 'temari' selections

There are some excellent choices among mophead, or in Japanese, temari forms of *H. serrata* such as **'Maiko'**, a delightful 0.3m (1ft) tiny shrub with pale-pink or mauve, small temari flowerheads, perfect for the smallest garden and that, planted in small groups, forms a charming carpet of pink ground-cover. Best in part-shade. It was originally found on Mt Daigo in Kyoto, only kept as a herbarium specimen, then rediscovered by Mr Yamamoto.

'Maiko' is a term for a young apprentice geisha, who sings and dances at banquets and parties but who does not yet possess the full set of geisha skills, like interesting conversation. **'Shiro Maiko'** was found in the foothills of Mt Fuji and is a white version of 'Maiko', which is more vigorous and less free-flowering, but attractive if grown with 'Maiko' as a companion planting.

Hydrangea serrata **'Besshi Temari'** is one of the best serrata mopheads, incredibly free with its pink or pale-blue heads that are crowded with shapely small flowers. It makes a small, domed shrub simply covered in flowers, which gradually fade to a creamy colour over several months. With moisture at the root it will tolerate some sun. It was cultivated since the Edo period,

'Maiko'.

'Besshi Temari'.

'Iyo Temari'.

then lost and like other valuable cultivars, rediscovered by Mr Yamamoto. Highly recommended.

'Iyo Temari' is a similar pale-pink temari cultivar, but with larger individual florets, shown here in its autumn colour.

There are a few excellent bicoloured flowers, the most popular and the most effective in the garden being the robust and vigorous picotee *H. serrata* **'Kiyosumi Sawa'**, which grows to 1.5m (5ft). It has white flowers strikingly edged with pink, or mauve blue in acid conditions. If grown from seed, the seedlings will show a variety of picotee flowers, all of some garden value. The habit is upright and rather stiff and with time, untidily

'Kiyosumi Sawa'.

'Tessa'.

loose. The foliage is a major feature, dark red/bronze on unfolding. It was found below the Kiyosumi mountain ridge in Chiba Prefecture. *Hydrangea serrata* **'Tessa'** is a pretty seedling from Kiyosumi, and typical of what may be expected from open-pollinated seed.

'Niji' means rainbow, from Shikoku. It is a small plant, under 1m (3ft), with quite large, matt-pink flowers blue at the centre, a suffusion like a loose watercolour. It is excellent for a patio or pot plant, to be enjoyed at close quarters. A more colourful White House farm seedling enhances this impact.

'Tosa-No-Ataksuki' means dawn in Tosa, an old name for Kochi Prefecture in Shikoku. It is quite distinct, the immature flowers a changing lively palette of green and red, less than 1m (3ft) in height and spread. A striking colour combination, if shortlived, transitioning to red as the flowers mature.

'Ohmachi Ezo' has almost circular, pink ray flowers set off by bright blue fertile flowers. The ray flowers never turn blue and both the shape of the ray flowers and the colour contrast are striking. The shrub is a typical serrata, about 1 × 1m (3 × 3ft) and rather upright; better in some shade.

Niji (left) and similar White House Farm *H. serrata* seedling (right).

'Tosa-No-Ataksuki'.

'Ohmachi Ezo' in neutral soil.

Variegated forms

'Kujuusan' is a rare variegated, small Japanese serrata, no more than 0.5 × 0.75m (1½ × 2½ft) across, a delicate-looking small shrub with pale-green foliage flecked and dusted with white. Critics claim it looks like a victim of red spider, but when bedecked with pale-blue lacecaps it has a unique quality. It seems to me less effective when the flowers are pink, as the blue tends to harmonise more effectively with the variegation. It gets its name from Mt Kuju in Oita Prefecture. Another variegated serrata, *H. serrata* 'Fuji Waterfall' AGM was considered a novelty for its variegation, perhaps a key factor behind its AGM.

'Kujuusan'.

Putative hybrid forms of *Hydrangea serrata*

There are four further plants that warrant description in this section, though they are almost certainly hybrids with *H. macrophylla*.

'Blue Deckle', raised by Michael Haworth-Booth, is a free-flowering, compact shrub with attractive light-blue or pale-pink flowers with a prominent deckle edge. Easy going and tolerant of a wide range of conditions, including full sun.

'Shojo', raised in Japan in 1992, grows to 1.5 × 1.5m (5 × 5ft) in part-shade, with foliage intermediate in size between the two parents and exquisitely shaped, rounded ray flowers, some 3.5cm (1½in) across, of a soft blue, with blue fertile flowers in lacecap style. I find it less appealing as a soft pink in alkaline soil, but as a blue it is in the top class.

'Odoriko Amacha' is another hybrid cultivar raised in Japan in 1960, a plant of similar growth with flowers freely borne in

'Blue Deckle'.

'Shojo'.

'Odoriko Amacha' (top and below).

'Warabe'.

***Hydrangea serrata* Forms Currently with the Award of Garden Merit, Awarded from Round Table Discussion by an RHS Panel of *Hydrangea* Cognoscenti**

- Diadem
- Miranda
- Bluebird
- Preziosa
- Rosalba
- Tiara
- Shiro Fuji
- Shojo
- Beni Yama
- Miyama Yae Murasaki
- Kiyosumi
- Grayswood
- Fuji No Taki
- Fuji Waterfall

early July, showing a striking contrast between the gentian-blue fertile flowers and pure-white ray flowers, tiered across the plant in a flat lacecap format. It will tolerate a good deal of sun, though does best in part-shade. It would be one of my first choices as an easy-going garden plant.

Hydrangea serrata **'Warabe'** is a charming pure-white mop-head macrophylla/serrata hybrid, also raised in Japan, with smaller heads of flowers that do not weigh down the bush. It is a neat plant, about 1 × 1m (3 × 3ft) spreading quite elegantly down to the ground; again, intermediate between the macrophylla and serrata parents. Ideal for a large pot, it matures to a soft pink in the autumn, which persists into winter.

White House Farm seedlings

I have been growing serratas from seed for some twenty years. My main objective in creating and selecting new hybrids was to develop a first-class, hardy, free-flowering, compact gentian-blue. I have achieved this with a number of cultivars, as yet not registered or on the market. They also make excellent rich reds in alkaline soils.

Open-pollinated seed has also yielded plants as good as, or an improvement on, many of those in cultivation. Although some are of obvious commercial value, these seedlings have not gained much interest in the trade, possibly because the market is not as big as that for macrophyllas, with plenty of effective promotion for big mopheads but an almost complete absence of promotion for serratas. Demand is producer-driven rather than based on the intrinsic merits of the product, about

A promising White House Farm seedling in its three modes: an excellent blue in acid soil (top); in neutral soil, a transitional mixture (middle); and in alkaline conditions, a rich red (bottom).

White House Farm *serrata* seedlings showing their free flowering capacity, able to compete for show with any macrophylla.

This page and opposite: A dozen blue *H. serrata* hybrids at White House Farm, showing new potential colour range.

which little is known by the general public, because no one is talking to them. This becomes clear with visiting groups, always enthusiastic about the plants and asking where they can be sourced. It is the classic Catch 22 situation – no trade promotion because of lack of demand and no demand because of lack of trade promotion. The status quo prevails.

Because they look more delicate, there may also be a deep-seated misconception among growers that serratas are difficult and fastidious plants of doubtful winter-hardiness that are not for every garden. This could not be further from the truth. See my comments on plants in Nova Scotia, where they perform beautifully after winters down to –20°C – for example, *H. serrata* 'Bluebird' (earlier in this chapter) flowers beautifully unperturbed by Nova Scotia winters.

On top of this, commercially, serratas are also slower to make a saleable plant.

The lack of trade interest is odd, because these plants will all flower comfortably in a pot and members of the public browsing in a garden centre will always buy something sporting a pretty flower. Until well-known, *H. serrata* needs active selling, especially those with unfamiliar Japanese names.

Most of the serratas I have bred at White House Farm show a free-flowering capability that would compete comfortably with any macrophylla in terms of garden impact. The plants shown are grown for the most part in full light with afternoon shade.

Hydrangea stylosa – the only widespread species in the Macrophyllae section not in Japan

Hydrangea stylosa is the only species within the Macrophyllae subgroup found outside Japan. It has a wide distribution through the eastern Himalaya, Sikkim and Bhutan, into Burma and west and south China, and down into North Vietnam, where it was known as *H. indochinensis* before being sunk into *H. macrophylla* ssp. *stylosa* by McClintock, who did not recognise *H. stylosa* at species level. The foliage has something of the thickness of *H. macrophylla* but the leaves are ovate, smaller and narrower, with length twice to four times as long as the width, 7–14cm (3–6in) long by 2.5–5cm (1–2in) wide. The fertile flowers are small, lavender, but often seen in a good form as a vivid gentian-blue, wearing a coronet of pure-white ray flowers with a conspicuous serrate deckle edge. Flowers are freely borne on a shrub about 1.5 × 1.5 m (5 × 5ft) upright and, with time, filling out as a spreading open shrub. In flower in June it is a plant of great distinction, especially with some rest from the midsummer sun in the afternoon. Plants raised from seed collected in western Yunnan have been undamaged by either cold or heat in over twenty years in Kent. This is yet another little-known hydrangea with potentially high garden value.

This page and opposite: *Hydrangea stylosa* has vivid-blue fertile flowers in good forms, with a conspicuous deckle edge to the ray flowers.

CHAPTER 7

PETALANTHAE: *HYDRANGEA SCANDENS* AND *HYDRANGEA CHINENSIS*

There are two species in the *Petalanthae* subsection, *H. hirta* and *H. scandens* (McClintock), but since the McClintock monograph *H. chinensis* has been broken out and is now generally accepted as a third species.

Walking the woods of Japan

Many Japanese hydrangea enthusiasts spend their leisure hours walking species-rich woodland, looking for particularly attractive plants or those with unconformities – anomalous variations and variegations, seedlings or sports of the species. A memorable experience that illustrates this woodland walking perfectly was when we were private guests of a retired Japanese schoolmaster, who showed us his collection of special botanical treasures, most of which he had found walking the woods in the school holidays. He had set them up for display in pots on shelves and chairs. There was one hydrangea that really took my eye, an exquisite dwarf form of *H. scandens*, with double flowers in extraordinary profusion.

Hydrangea scandens 'Yoshiko Yae' showing scandens habit flowering along the shoot from the leaf axils (left) and its flower power (right).

OPPOSITE PAGE: Two contrasting specimens of *H. chinensis* from Taiwan showing the typical yellow fertile flowers, opening up new landscape colour possibilities for the garden; a garden-worthy form (top) and *H. chinensis* forma *macrosepala* (bottom).

I kept returning to this superlative plant, photographing, looking at and noting the detail, in clear admiration. He came up to me and explained that he had found it in the forest on one of his walks and asked if I would like cuttings. He then said 'If you and I are the only people in the whole world to have this plant we should give it a name. I think we should name it after my wife, Yoshiko, and because it is double it should be called **"Yoshiko Yae"**.' His wife, listening in the background, exquisitely dressed in what I took to be a traditional kimono in embroidered gold, gently bowed in acknowledgement.

Taxonomic change

McClintock treated *H. chinensis* as a subspecies of the Japanese *H. scandens* but it is now accepted as a distinct species. It does not occur in Japan, but is widespread elsewhere in East Asia, from western and southern China to Vietnam, Burma, Taiwan and the Philippines. Like *H. aspera*, with a wide distribution, it is highly variable, to the point where McClintock sank no less than 24 taxa into what she designated as *H. scandens* ssp. *chinensis*, all previously regarded as species in their own right. Some of these sunk species are again today being recognised at species level and more than a few have notable horticultural merit.

The essential differences between *H. chinensis* and *H. scandens* are threefold: *H. chinensis* generally makes a much bigger shrub, sometimes reaching up to 3m (9ft) in cultivation, while *H. scandens* is a spreading, rather low plant of modest size, usually about 1m x 1m (3 x 3ft); second, the *H. chinensis* leaves are altogether larger, with length normally being some two to three times as long as their width, but occasionally narrower and lanceolate, and with intermediate forms to be found in the same population. *H. scandens* has smaller leaves of rather thin, papery texture. The third difference is that *H. chinensis* flowers terminally, while *H. scandens* also bears flowers in the leaf axils, along the stem. This is an important character for discerning hybridists, and likely to play a significant part in future garden hybrid production.

Three distinctive foliage forms of *H. chinensis* collected by Crug Farm Plants in Vietnam, all collected in close proximity in the same wild community.

Exploring the area around Huidong in southern Sichuan just north of the Yangste in 1994, we came across a considerable acreage of old abandoned paddy terracing that had reverted to the wild and been taken over mainly by an extensive duopoly of *H. chinensis* and a *Hypericum* species, forming an almost continuous blanket about 1–1.2m (3–3¾ft) high. As it was October there were no fresh flowers, but the hydrangeas still bore some faded and dried ray flowers: I collected seed from a specimen with unusually large ray flowers. Subsequently, most of the handful of seedlings that I raised and kept to grow-on to flowering size reverted to 'normal' sized flowers, but one turned out to be as large, if not larger than, the original parent plant. I labelled this temporarily and descriptively for my own purposes **'Big White'** and gave away cuttings to friends. Though requested to name it properly, I failed to do so and the original label note of 'Big White' has now stuck. I notice it has

Hydrangea chinensis 'Big White'.

A dark blue form of *H. chinensis* from Crug Farm Plants.

trickled down and out, and is now on offer in the RHS Plant Finder from two nurseries in the UK and probably more in Europe, some 28 years on from its raising. It is a typical form of Chinese *H. chinensis* but its flower size and substance make it a rather effective garden plant, with four to five white ray flowers up to 6cm (2in) across, surrounding a rather open dome of white fertile flowers, the whole lasting into autumn, fading to a greenish buff. It also has larger leaves up to 15cm (6in) long by 5cm (2in) wide with a pronounced acuminate tip. It grows to about 1.2m (3¾ft) in partial shade and has turned out to be perfectly hardy in Kent. It needs good drainage, as do all these typical *H. chinensis* forms.

I have another unnamed plant with even bigger creamy-white ray flowers, up to 8cm (3in) across, with as many as seven rays to a corymb, surrounding a sparse, open series of rather few, fertile flowers. The foliage is also large, a papery oblong/ovate, sharply serrate leaf twice as long as wide, some 13 × 7cm (5 × 3in). The plant grows in shade until early afternoon.

An unnamed large-flowered form of *H. chinensis*.

The point is that *H. chinensis* is very variable, and some of its best forms are eminently gardenworthy, but simply unknown in the trade, which shows little inclination to take them on board. They will never rival in colour impact the big fat hortensias, but there are many gardeners who appreciate a more subtle, less 'in-yer-face' effect, which may be consistent with the more informal style of their garden.

McClintock based her revision on dried herbarium material, which does not show flower colour, but she concluded from the occasional collector's label that *H. chinensis* fertile flowers could vary from white to yellow to blue. This is borne out by various collections over recent years, some with fertile flowers of a rich purple/blue and others of a conspicuous yellow. A classic example of outstanding quality is a blue form collected in Vietnam by the Crug team of Bleddyn and Sue Wynn-Jones. It has fertile flowers of the most intense purple/blue, with white ray flowers, to create a stunning effect.

The Taiwanese *Hydrangea chinensis* introductions

Most Chinese forms have white-on-white fertile and ray flowers, but a whole raft of *H. chinensis* forms have been introduced by the Wynne-Joneses from Taiwan. The best specimens of these are of significant garden value and for the most part bear yellow fertile flowers in contrast to the white ray flowers. They are semi-evergreen in a mild winter.

The Taiwanese forms are generally distinct from the Chinese forms, too, for their smaller dark-green, lanceolate leaves and

Some forms are worth a place in the garden.

A typically good form of a Taiwanese collection of *H. chinensis*.

Hydrangea chinensis forma *angustipetala*.

Hydrangea chinensis forma *formosana*.

smaller flowers. McClintock placed them all as named subspecies of *H. scandens*, but some authorities have raised them to species level, while the most recent revision places them as forms of the newly recognised *H. chinensis*. Several forms have been named, such as forma *angustipetala*, *formosana*, *macrosepala* and *obovatifolia*. These plants are so far the nearest thing to a yellow hydrangea, all having eye-catching, yellow, fertile flowers, some a deep rich colour.

Hydrangea chinensis forma ***angustipetala*** is variable (Crug lists nine collections) but some are worth a place in the garden. Their foliage is light and elegant, for the most part willow-like, some 13cm (5in) long by 2.5cm (1in) wide, with a long tapering drip tip. They will reach up to 3m (9ft) high in cultivation, erect, branching freely from the base, spreading at maturity: this after ten years, without fertiliser and though enjoying a moist soil, with irrigation only *in extremis*.

The 3.5cm (1½in) white ray flowers are in an informal ring of up to a dozen, crowding around and into a bright-yellow dome of fertile flowers. I grow a distinct form from a Crug collection with fewer, smaller ray flowers, but a big tumbling mass of small creamy-yellow fertile flowers, a bit like meadow sweet, with a pleasant fragrance. These fade to a pale green in the autumn. The whole makes a pleasing picture in the garden landscape, seen at its best and lasting longest in partial shade, whether from a tree or large shrub or against a fence or wall facing north.

A pale blue form of *H. chinensis*.

Seedlings from forms of *H. chinensis* forma ***formosana*** are variable in flower, some with bright-yellow fertile flowers.

The largest ray flowers are borne by forma ***macrosepala*** – the name is the give-away – and individual flowers, usually with only three or four florets, can measure as much as 6–7cm (2½in) across, or twice the size of, say, a typical forma *angustipetala*. These can be as many as ten to a corymb, but usually fewer, irregularly notched, informally decorating the yellow fertile flowers. The habit and foliage are indistinguishable from forma *angustipetala*, but the flowers are noteworthily large and certainly distinguished.

Finally, there is *H. chinensis* forma ***obovatifolia***: some seedlings exude charm, with ray flowers virtually circular, the individual florets overlapping to create a refined and elegant shape. The ray flowers are very freely borne, scattered randomly across a quite open and small corymb of fertile flowers, which are a pale greenish-yellow. This a delightful plant to grace any garden.

Hydrangea chinensis forma *obovatifolia*: a Crug seedling, an excellent form, which grows just opposite the nursery office. A charming hydrangea, but rarely seen in gardens.

Japanese *Hydrangea scandens*

Hydrangea scandens is a Japanese species, otherwise found only rarely on Taiwan. It has not been found in China. It makes a rather loose shrub in part-shade, up to 1.5m (5ft), with small, papery, matt leaves some three times as long as wide, from 3 to

Hydrangea scandens in deep woodland with Bleddyn Wynn Jones of Crug Farm Plants.

H. chinensis with green – and fragrant – fertile flowers.

Hydrangea luteo-venosa, blooms in April, the first hydrangea to flower.

8cm (1–3in) long by 1–3cm (½–1in) wide. The pale-yellow fertile flowers are generally few, on loosely branching pedicels, with a less than generous scattering of white ray flowers.

There appears to be another definable form that in good growing conditions produces long, semi-scandent shoots that flower from end to end the following year, with corymbs arising from the leaf axils. The weight of these flowers pulls down the scandent shoots to create a floriferous, graceful, arching garland. This is the most notable character of the species and of interest to hybridists, who might also note its early flowering, beginning in April.

Apart from *H. scandens* ssp. *chinensis*, now accepted as a full species in its own right as *H. chinensis*, McClintock recognised one other subspecies, *H. scandens* ssp. *liukiuensis*, into which she merged *H. luteo-venosa* (Koidzumi, 1925). These two plants have an uneasy taxonomic relationship, with some authorities treating the latter as a form, others as a full species. Both are close allies of *H. scandens*, and Bleddyn Wynn-Jones reports both *H. scandens* and *H. luteo-venosa* arising from the same seed batch. They are quite similar in the garden, with a few typically small leaves occasionally randomly streaked and freckled with creamy-yellowish variegation. There seems to be less variegation, and that mostly white, when in shade, which the plants prefer, if they are to grow and flower optimally. I have found *H. scandens* ssp. *liukiuensis* difficult to establish and slow to get going, a weak twiggy thing, and uncooperative, eventually reaching some 0.5m (1½ft) by as much across. By comparison, *H. luteo-venosa*, from a commercial source, has made a shrub in partial shade to about 1.5m (5ft) with a similar spread after 25 years. As it matures it flowers very freely and, because it is very early, beginning to flower in April, it is effective in the garden landscape and a lighter touch than the spring heavyweights camellia, magnolia and rhododendron. It is the very first hydrangea to flower, an *amuse-bouche* to sharpen the appetite for the coming summer banquet.

The French missionary and plant collector Abbé David discovered *Hydrangea davidii* in 1869 on the hillsides near his mission at Baoxing in Sichuan: it was named by Franchet to commemorate him. (*H.* **davidii** was sunk by McClintock into *H. scandens* ssp. *chinensis* but is now recognised at species level.) It makes a shrub of about 1m (3ft) with ovate/lanceolate leaves, slightly recurved, dark-green above, with tufts of hair in the axils of the veins on the lower side. The flowers are distinctive, the fertiles dark blue in sparse, open corymbs on purplish pedicels, and the white ray flowers often with three petals rather than four, the lower one bigger, so that it sometimes gives the impression of a pansy flower or a Mickey Mouse face. I find it difficult to be sure of distinguishing it from forms of *H. chinensis.*

Although it is very ornamental, *H. davidii* is rare in gardens, perhaps because it is difficult to establish and proves to be an uncooperatively recalcitrant little plant. I have lost four accessions from three different collectors, each establishing encouragingly at first, then after two to three years dying back

Hydrangea davidii, now once again recognised as a species.

and gradually dwindling to a disappointing and frustrating end from no obvious cause (in spite of other species nearby in the same conditions growing lustily).

Hydrangea hirta is the only species among all the Asians that is notable for having no sterile ray flowers. It is quite frequent in Honshu in Japan but restricted to that province, and occurs elsewhere only in the Ryukyu islands, on Okinawa. Botanically, in fertile flower and fruit, it is close to *H. scandens*, but differs in two distinctive characters: its lack of ray flowers, plus a boldly serrate leaf, closely resembling a nettle. It makes a dense, twiggy shrub of about a metre across and high, with small corymbs of RAF blue, fertile flowers. It is an interesting collector's item, of little garden value. There is a rare white form.

I was given a quietly attractive, naturally occurring hybrid, *H. hirta* crossed either with *H. scandens* or *H. serrata*, which has pretty, scattered, mid-blue ray flowers with rounded tepals, often occurring singly or in twos in some of the fertile flowered clusters. It makes a nice low, twiggy bush in part-shade, with a suggestion of the nettle-like foliage of *H. hirta* and about every fourth corymb bearing one or two pretty blue ray flowers.

A naturally occurring hybrid of *H. hirta* and *H. serrata*, but lacking the bold nettle-like foliage of *H. hirta*.

CHAPTER 8

CALYPTRANTHAE: THE HARDY CLIMBERS

According to the McLintock treatment, there is only one species in this subsection, namely *H. anomala.* This comprises two subspecies – ssp. *anomala*, the type, and ssp. *petiolaris*. The only botanical difference lies in the number of stamens. The McLintock identifying key to the subspecies is perhaps the shortest on record, namely:

1 Stamens 9–15..........ssp. *anomola*
1 Stamens 15–20........ssp. *petiolaris*

That's it in botanical terms; but as usual in Hydrangeaceae, there is some taxonomic confusion, with half the botanists following the McLintock subspecies formula, and the other half according full species status to *H. petiolaris*. However, in horticultural terms the differences are significant, with *H. petiolaris* the better of the two for garden landscape effect and general planting, with more and bigger flower corymbs with more prominent ray flowers and reliable hardiness. Having said that, there are one or two recent introductions of *H. anomala* that should also be considered, notably from the Crug Farm team who have introduced so many fine species from the wild.

Both species are vigorous self-clinging climbers that behave precisely like common English ivy, with rapidly growing climbing shoots fastened firmly with aerial roots into any support; and shoots with nothing to climb forming a bushy mass of flowering laterals. They can readily swallow a small building whole in a few seasons and in a tree reach up to 20m (60ft).

Climbing hydrangeas in the wild and pink variants

Hydrangea anomala is widespread across the eastern Himalaya, north India, and west and central China. It is found from Nepal through Sikkim and Bhutan, Uttar Pradesh and Himachal Pradesh to south-eastern Tibet and in China from Yunnan to Hunan. In China it has an extensive altitude range, from a low 600m (1,920ft) to an alpine 3,800m (12,160ft). In the Himalaya it ranges from 1,200 to 2,500m (3,840–8,000ft) according to McLintock, though I believe the Wynne-Joneses may have found it lower, around Darjeeling in India, when it might lack full hardiness in cultivation.

Hydrangea petiolaris is Japanese, from Hakkaido in the north right down to the island of Yakushima in the far south. It also occurs in Russia (Sakhalin), Taiwan and the Korean islands of Cheju do and Ullung do.

The flowers of ***H. petiolaris*** are a creamy white, with the fertile flowers a matt ivory and up to six clean, pure-white ray flowers per inflorescence some 4cm (1½in) across. The quality

OPPOSITE PAGE: This is a highly unusual pink form of *H. anomola* collected by Crug Farm Plants in Taiwan and known as 'Crug Coral'. A fast, self-clinging tree climber.

Hydrangea petiolaris from Yakushima in Japan, flowers, leaf reverse and attractive young growth.

and size of the flower is variable, but the general effect in a good form is of a loose, lacy veil some 15–20cm (6–8in) across, at its best in late June into July, but fading slowly and still effective into autumn.

Hydrangea anomala is also white, but with smaller and usually fewer ray flowers, which in some forms are entirely absent. There are, however, attractive pink forms, and Crug Farm have introduced a form they have dubbed *H. anomala* ssp. *glabra* **'Crug Coral'**, which produces coral-pink ray flowers. They found it in an isolated colony in the central mountains of Taiwan. *Hydrangea glabra* was classified by Hayata as a species, but was sunk by McLintock into *H. anomala* ssp. *petiolaris*. She did not see the type specimen (Kawakami and Mori), so the 'glabra' monicker may be doubtful.

Other good pink forms of *H. anomola* do exist in the wild, but as yet, apart from 'Crug Coral', remain to be introduced. I was in panda country in Sichuan in May 1998 to see and photograph spring flowers, and saw a plant of *H. anomala* in the forest with relatively small but lovely deep-pink ray flowers. The seed of the previous year had evidently all been shaken out by boisterous winter winds but I took a couple of old flower-heads, crunched them up, then sowed the debris. Big excitement

Hydrangea anomala ssp. *glabra* 'Crug Coral'.

when just two seedlings emerged and grew on nicely. Both flowered in year five. Both had white ray flowers. *C'est la vie*.

Climbing hydrangeas for the garden

While *H. anomala* is rarely encountered in cultivation, *H. petiolaris* is now widely planted, and can be commonly seen growing freely on north walls where, in spite of the relative lack of sun, it still contrives to flower well.

A note of caution here. If contemplating house-wall planting, you should be aware that the adventitious aerial roots cling

Hydrangea petiolaris smothering a high wall.

H. anomola (left) is generally no match for *H. petiolaris* – especially this good form from Yakushima, Japan (right)

tenaciously to the wall. Should you wish, for any reason, to get rid of or cut back the plant, they obstinately refuse to go and bits remain as a tribute to its tenacity, and are difficult to remove. Furthermore, if you have painted white walls, as we do at White House Farm, and wish to pull off the climbing shoots, their roots will remove large flakes of paint or plaster, defacing the wall like the dark tracks of some giant centipede. Although *H. petiolaris* can be used to cover effectively any unsightly object, such as an old tree stump or an old shed or barn, perhaps its most impressive performance is achieved by allowing it to climb a suitable tree. Generally, many of the seedlings introduced over recent years by Crug are superior in flower and foliage to that normally available in garden centres and we grow one of their first-rate introductions from Yakushima into a sweet chestnut tree, limbed up to a clean trunk of around 8–9m (24–27ft).

Here it thrives, producing short lateral shoots that ring the trunk with flower. The species is often raised from seed and to make sure it is of acceptable garden quality – good size and number of ray flowers – I initially prevent its climbing, holding it back as a shrub until it produces its first flowers, usually after four or five years. If it gets the thumbs up as a good form, I allow it to enjoy the freedom of the tree; because it is already firmly fixed and given a start, it will streak north and comfortably make a good 2m (6ft) growth in a year. It could eventually reach to around 20m (60ft). It needs some moisture to grow well and is best planted on the north side of the tree in the shade, allowing it to grow into the sun on the south side where it flowers very freely.

An alternative use for these plants is to grow them as a free-standing shrub, fully out in the open, where they make a dense, twiggy bush, covered in lacy flowers. Usually seen up to about 1.5m (5ft) and perhaps twice as much across, the bush is easily contained within required boundaries by judicious pruning and clipping, and it remains exceptionally free-flowering.

A few cultivars of climbing hydrangeas

Apart from 'Crug Coral', I have grown only two other petiolaris cultivars over the years, namely 'Cordifolia' and 'Brookside

Seen in Japan, a new variegated form of *H. petiolaris*.

Littleleaf', both 'miniature' forms of *H. petiolaris*. These are so similar they could be the same plant, or confused in the trade so that in buying both, you could be really buying one and the same thing. They are both, in effect, smaller versions of the species, with neat, small, serrate and, as the name suggests, heart-shaped leaves. That's about it, really. I grew them on the side of some steps in a favourable spot in part-shade but with good light, and they produced plenty of foliage but few or indeed no flowers over a number of years. They failed to pay their rent so were eventually discarded to the bonfire. They may do better on a warm south wall.

Crug have named a form of *H. anomala* that they collected in Sikkim as 'Winter Glow'. They describe it on their website as an evergreen with them in North Wales, with serrulate, thin, textured foliage that turns purple over winter. It evidently grows in a fairly exposed situation on an east wall with green, domed, lacy flowers. There are quite a few variegated sorts available for those who appreciate these anomalous forms, mainly originating in Japan and the USA. These can be viewed online on a variety of websites. *Hydrangea anomala* 'Kuga Variegated', discovered as a branch sport in Japan, is protected by patent. It has young white, yellow and pink shoots transitioning through to speckled and splashed leaves which become green at maturity. *Hydrangea petiolaris* 'Silver Lining' is another protected variety, which is generally available with greyish-green cordate leaves with a silver edge and typical *H. petiolaris* flowers and vigour.

There are other variegated forms available for those keen on variegation and while in Japan, where these streaked and mottled plants are popular and sought after, I saw various small variegated plants featured in collections in pots, some without names, so there could be further variegated introductions in the pipeline. Growers in the USA are also on the *qui vive* for anomalous forms.

CHAPTER 9

THE CORNIDIA SECTION: SUBTROPICAL LIANAS

These are evergreen climbing hydrangeas, also ascending by aerial roots, like ivy. McClintock lists twelve species in two subsections. They are native to Mexico, Central and South America, with just one species, *H. integrifolia*, in Asia, found in the Philippines and Taiwan. This is hardy enough, as is the single species in Mexico, *H. seemannii*, but the majority are subtropical and, therefore, of only marginal interest for gardeners, as they are not sufficiently hardy to survive cold temperate conditions, though one or two of the borderline species might now be worth a trial given our recent reasonably prolonged run of relatively mild winters and the forecast of more to come.

The two subsections in Cornidia are Monosegia and Polysegia. These differ botanically only in their inflorescence: the former has single, terminal, flower cymes, the latter multiple, as the names suggest. There are two reliably hardy species in the Monosegia subsection, namely *H. integrifolia* and *H. seemannii*. Even though they are geographically separate, one from Asia and the other from the Americas, they are very similar insofar as both their fertile and ray flowers are white. Like *H. involucrata* forms in Japan, these plants have large, rounded, involucre-wrapped buds. The only other species with white flowers in the Monosegia subsection is *H. asterolasia*, which is found in Costa Rica, Panama, Columbia and Ecuador, and so is not hardy.

Hydrangea integrifolia from Taiwan growing on a massive vertical cliff.

OPPOSITE PAGE: This dramatic bright-red member of the Cornidia section, called *H. trianae*, climbs trees in Columbia where it was photographed by Bleddyn and Sue Wynn-Jones of Crug Farm Plants. It is not in cultivation as yet, and will not be hardy, but its potential as a parent crossed with hardy kinds could mean a colour breakthrough in climbing hydrangeas.

Hydrangea seemannii covering the roof of an outbuilding at Crug Farm Plants (top). The same plant showing flower detail (middle) and in bud (bottom).

Hydrangea serratifolia arrangement showing flower.

By contrast, *H. integrifolia* is found at altitudes of up to 3,000m (9,000ft) in the central mountains of Taiwan, where Bleddyn and Sue Wynne-Jones describe it as growing vigorously on steep cliffs, as well as into trees, and regard it as easily the hardiest of the evergreen climbers. The leaves are ovate, up to 18cm (7in) long, margins entire with some remote serration, dark, shiny green. The conspicuous, perfectly round flower buds, contained in wrap-around bracts (involucres) crack and burst, and a froth of white fertile flowers begins to emerge, eventually forming a lacecap-style flower, a tumble of small fertile flowers scattered randomly with white ray flowers. It will grow comfortably to great heights into a tree, or it can be grown as a free-standing bush, suitably pruned; it can also make an impressive tall, clipped hedge.

Hydrangea seemannii, though from far-off Mexico, is very similar to *H. integrifolia* in most characters of foliage and flower, but variable from seed, some seedlings with relatively small and inconspicuous ray flowers. Others are much more impressive as garden plants, with larger and more numerous ray flowers, which on a warm wall or on a south-facing roof of a low building can produce a tumbling turbulence of white and cream to rival any other climber as an effective garden plant. It needs full light with moisture at the root to flower really freely. In a good selected form, the ray flowers are shapely, relatively large at 3cm (1in) plus across and with overlapping petals produce an almost circular flower; these are pure white on long, thin pedicels and surround a mass of cream fertile flowers. The key is to find a good form.

Hydrangea serratifolia from central Chile is the type species of the Polysegia subsection and the only other Cornidia section hydrangea with a reasonable claim to be hardy. I have it in Kent topping out a 20m (60ft) oak twelve years from planting. The name is a misnomer, the leaves having entire margins, that is smooth edged without serrations, or, as the botanists say, remotely serrate; you can find the occasional leaf with a suggestion of serrations if you look hard enough. (It used to be called, rather more accurately, *H. integerrima*, meaning 'entire' or without serrations.) It will also run along the ground, rooting as it goes and, if unchecked, can seriously overwhelm neighbouring plants. If there is adequate space to allow it freedom, it makes effective ground-cover. Leaves are, typically, up to about 14cm (5½in), ovate, dark, glossy green and two to three times as long as wide. It is not spectacular in flower, all the flowers being of the small fertile type, mostly without ray flowers but with prominent stamens.

Hydrangea peruviana × *H. integrifolia* draping a tree (left). Its pink flowers without ray flowers (right).

No other members of the Cornidia section are reliably hardy. This is our loss, because McClintock quotes Standley (author of the *Flora of Costa Rica*) referring to *H. oerstedii* as '...one of the most beautiful plants of Costa Rica because of its broad cymes of showy flowers which are not inferior in beauty to those of cultivated species'. Both fertile and ray flowers are rich pink. *Hydrangea oerstedii* is botanically closely allied to *H. peruviana*, the difference distinguished only by the relative length of the stamens and styles. But all is not lost, in that there is a hybrid between *H. peruviana* and *H. integrifolia* with bright-pink fertile flowers, although without ray flowers, that appears to be at least moderately hardy, as it is up some 18–20m (54–60ft) into a tree at Tregrehan, Tom Hudson's celebrated garden on the south coast of Cornwall, where it flowers freely in part-shade. We are trying it in Kent.

So there are reasons to be cautiously optimistic, both because the cross is compatible, certainly within the Cornidia section, and also because it seems that both the pink colour and relative hardiness may be dominant characters of such hybrids. There is thus a real possibility of a breakthrough with glorious red/pink ray flowers from deliberately crossing these great climbers. This is further discussed in Chapter 13 on future developments, especially with regard to the spectacular red-flowered Crug discovery of *H. trianae* in Columbia (the image that opens this chapter).

Inside the involucre of the pink *H. peruviana* × *H. integrifolia* at Tregrehan, Cornwall, where it is hardy.

PART III PRACTICALITIES

CHAPTER 10

USES AND GOOD PRACTICE IN THE GARDEN

The mixed border

After June, the fabulous spring colours in the garden have disappeared. Summer colour is at a premium, and is typically provided by single, isolated major features such as the celebrated English herbaceous border, formal patterns of colourful annuals or formal beds of roses. The herbaceous border, which was essentially a Victorian invention, can indeed be a real spectacle, a dramatic riot of colour, offering rafts of perennial plants typically arranged in complementary or contrasting colour-coded sections, with big plants at the back and little plants in the front. Its basic *raison d'être* is colour, the more the merrier and the bigger the scale the better. However, the full monty herbaceous border is seriously labour-intensive and very high-maintenance: staking, lifting, dividing, planting, manuring, overwintering, weeding, trimming, deadheading – it needs an army of gardeners to do it full justice. Due mainly to its high labour costs, it has rather fallen out of fashion over recent years in favour of the cheaper, lower maintenance mixed border with shrubs and other more easily maintained plants, like grasses, being incorporated.

This is where the hydrangea is beginning to come into play, with its long season of colour from June until autumn frosts, and its relaxed, undemanding character, tolerant of most soils – and reliability, with virtually no maintenance required. *Hydrangea aspera* 'Macrophylla' is used in the Wisley herbaceous border to provide high contrast of colourful flowers and foliage, a successful backdrop to a host of colourful perennials. *Hydrangea paniculata* cultivars, pruned up, can also provide height and some summer shade protection for more delicate herbaceous plants, as well as weeks of flower. Apart from contributing substantially to colour in the mixed border, the hydrangea is unrivalled in providing masses of colour over summer months at negligible cost, and massed plantings of *H. macrophylla* cultivars provide just the eyeful needed to satisfy a summer public enjoying the sunshine. Colourful hydrangeas are beginning to seriously contribute to sustaining summer footfall for large gardens, which always need to raise visitor income and cut costs. Garden visiting is a growing popular leisure activity and many gardens are beginning to appreciate the role of the hydrangea in helping to bring people in.

Foundation plantings round buildings

As wall shrubs, too, hydrangeas can play a dominant role: a big spread or series of colourful *H. macrophylla* can enhance the profile of either small cottage or huge mansion, anchoring its summer presence in a solid bolster of green with months of

OPPOSITE PAGE: *Hydrangea involucrata* 'Chichibu' is a selected form of the wild species, with extra ray flowers informally arranged and attractively waved, introduced by The Ghent Botanic Garden in Belgium. *H. serrata*s make good cut flowers.

Hydrangea serrata softening the hard lines of Takahada Fudo Temple building in Tokyo.

colour. Other buildings can also be a canvas for colour, or simply have their edges softened.

Some village churches in Brittany are encircled and enhanced by masses of multicoloured mophead *H. macrophyllas* billowing along their walls like the sea. Climbing evergreen hydrangeas, if allowed, will happily swallow up smaller buildings and massive trees in a few seasons, and can become a stunning feature in bloom, and along with climbing roses and supporting clematis, a lesson in vertical gardening.

Woodland informality

Hydrangeas are ideal for planting informally at the edge of woodland, or to define the border of a multi-tiered bed. *Hydrangea aspera*, *arborescens*, *macrophylla*, *paniculata* and *quercifolia* cultivars, for example, define in summer the broad walk up into Battleston Hill at RHS Wisley, providing interest and spectacle for many weeks. All are also used to good effect when exploring the light woodland beyond the path, mixed with rhododendrons, camellias, magnolias and maples. Some can be used as small trees

Hydrangeas in an informal garden setting.

in this setting, others as light relief to big evergreens, complementing them with both foliage and flower, themselves enhanced by the evergreen backdrop: ideal companions.

Hedges

Some hydrangeas are used as effective internal hedges, such as cultivars of *H. serrata* flanking and defining a path. They are easily maintained, have no potentially aggressive characters like thorns or spiky leaves, and will provide a summer of flower if simply left unpruned and informal. I note that 'Annabelle' is being advertised precisely for such a 1m (3ft) hedging application and clippable without losing flower. As a dense but decorative screen, an unpruned informal hedge of heavy flowered cultivars of *H. paniculata* is also effective.

Contrasting companions

Hydrangeas are also excellent companions for large evergreens like camellias, rhododendrons, laurels and aucubas, supplying year-round contrast with their light foliage texture and offering eye-catching colour enhanced by the evergreen backdrop when its spring flowers are only a memory. The sometimes dark and overwhelming weight of these massive evergreens when out of flower – denizens of the old Victorian 'shrubbery' – motivated Reginald Farrer to write of hardy hybrid rhododendrons, 'At these repulsive pies, our offended gorges rise' (quoting Gilbert and Sullivan on pork pies). The potentially depressing effect of these heavies can be mitigated by massed colour of hortensia mopheads in a formal setting or lightened in a more informal environment by a generous fringe of low-growing, delicate and finely textured *H. serrata* cultivars that draw the eye and flower all summer, themselves benefiting from the dark backdrop – especially in various shades and tones of blue, that rare and desirable colour.

An informal mixed hedge of *H. serrata* seedlings.

Hydrangeas in an informal woodland setting.

Japanese *H. serratas* in light woodland.

When the aim is blue

Cool and relaxing, sea and sky, open spaces, freedom, serenity, tranquillity, imagination, the list of positives goes on – all life-giving attributes of the colour blue, according to psychologists. Blue is one of the most desirable colours sought and enjoyed in garden landscapes. And it is not a common colour, with only a handful of flowering plants – cornflowers, agapanthus, delphiniums, forget-me-nots, bluebells, hyacinths and a few more – providing pools of true blue in the garden. But the many different shades and tones of blue in the lexicon – azure, caerolean, cobalt, indigo, sky, sea, slate, navy, royal – are all available from one wonder plant: the hydrangea. It is without equal in providing blue over many weeks throughout the summer. It is a blue phenomenon in the right soils, or modified with soil treatment.

The colour blue is mainly confined, of course, to the two Macrophyllae species, *H. macrophylla* and *H. serrata*, which are the only species to provide blue *en masse*. Other species, such as *H. stylosa*, *H. aspera* and *H. chinensis*, can display vivid blue

Montage of *Hydrangea* blues across the genus.

A *serrata* hybrid, 'Odoriko Amacha' and *H. stylosa*. Both have vivid-blue fertile flowers.

Varying pH gives varying colour, including effects on the same plants.

fertile flowers, which can be effective in the garden, especially when contrasted with pure-white ray flowers. All the other hydrangea species are white (some with yellow fertile flowers) or pink, or a combination of these.

Perhaps the most frequent question asked of growers is: 'How can I grow blue hydrangeas?'

The basic chemistry providing blue flowers centres on the availability of free aluminium in the soil. Aluminium is 'mobile' in acid soils, and thus available to plants, but locked-up chemically and inaccessible to the plants in limy or 'basic' soils. The pigment providing the red colour in hydrangea sepals is an anthocyanin, known as delphinidin-3-glucoside. In acid soil conditions, as aluminium becomes incorporated into the hydrangea sepals, it ultimately forms a blue 'complex' with the anthocyanin. Unless you are a chemist with specialist knowledge, the detailed processes as to how this takes place on the molecular level are not relevant: suffice it to say, though, this makes sense of the fact that deep or pale colours tend to remain broadly in character, wherever the final tint lands on the blue–red spectrum (with all the purples and mauves in between). Other metals, such as tin, molybdenum or iron, can also impact colour change, through changing sensitivities to the soil pH, perhaps in combination with the aluminium.

The soil pH is the key element – below pH 6 and blue is easily achieved; for anything much above pH 7 or neutral, the best advice is to go with the flow and stick with the rewarding and wide choice of excellent reds, pinks and whites. If blue is a fundamental requirement in such conditions, building a raised bed and filling it with an appropriate acid compost (such as ericaceous) will be necessary. Soil values between pH 6 and 7 can also be adjusted to achieve good blues by adding aluminium to the soil in the form of aluminium sulphate. This is the effective ingredient in the proprietory 'blueing' powders available on the market. A bag bought from the local ironmonger is cheaper, and works equally well. It can also be used to intensify the blue in already acid soils. Modes of application and rates of use do rather vary in the literature, but the golden rule is that too much is worse than too little: slightly frustrating the blue objective is better than damaging the plant.

When I bought my first house, I was delighted to find large, mature hydrangeas on the north-west wall. Like all thoughtful new owners I did nothing to the garden, but waited to see what would happen when it awoke in the spring: the colour and quality of the flowering plants, and what bulbs, corms and herbaceous stools might be below the surface. The hydrangeas were a uniform icing-sugar pink, probably as a result of residual builders' rubble, which was disappointing as the natural soil pH in the area was about pH 6, and healthy rhododendrons were growing in nearby gardens. So I tried feeding them with aluminium sulphate by topdressing with generous handfuls the following autumn, impatient and thinking a large helping would speed the effect, with winter rains to take it down to the roots. The result was a slowly unfolding disaster, the plants so badly affected with dieback and checked growth that in the end I had to remove them and start again.

Adding aluminium sulphate can thus be a risky business; Haworth-Booth recommends a surface dressing in November and again in spring, with overall no more than 4.5kg (10lb) applied to a large established plant. This seems excessive after my experience: perhaps concentrations of product vary. As an alternative, he recommends 7g/4.5ltr (¼oz/gallon) of water in autumn and spring, continuing with watering weekly until flowering and ensuring every part of the root system is covered. I have not tried either method as I am happy with blues from my natural soil, so cannot endorse it from the benefit of practical personal experience.

The best indicator of a successful colour-change intervention, easily noticeable with the lacecaps, is the tiny fertile flowers at the centre of the inflorescence. These are particularly sensitive and change hue long before the ray flowers. Interestingly enough, I have two *H. serrata* seedlings with intensely blue fertile flowers, but the ray flowers have remained a deep red/pink over many years, even in good acid soil, seemingly a genetic anomaly. The old *H. serrata* variety **'Hagoromo'** is known by the French as **'Toujours Rose'** for the same reason.

As hydrangeas are happiest with adequate moisture at the root, it is often necessary to irrigate during warm spells in summer, so it is well worth checking the pH of your water supply. My own mains supply is from aquifers deep in the chalk of the Kent downs, and during a heatwave, had to use this water to saturate

Hydrangea serrata White House Farm seedling: fertile flowers can blue in acid conditions, while ray flowers remain in the red spectrum.

Hydrangea serrata 'Hagoromo' remains pink regardless of soil pH. In France it's known as 'Toujours Rose'.

Hydrangea 'Blaumeise' watered with alkaline mains water supply in Kent: before (left) and after (right).

Hydrangea macrophylla 'Westfalen', a bright red.

the magnificent blue *H. macrophylla* 'Blaumeise' and deepest purple 'Nachtigall', both badly affected by drought. The result was that the striking blue/purple colours turned red/pink, and their former colour did not reappear for three years, gradually transitioning back to blue each year. The same transition from sky blue to pink when mains-watered, also occurred in my *H. serrata* 'Tiara'; it, too, has taken three years to revert to its former blue.

If the intensity of the blue desired is hard to achieve, or has been recently lost, another cause can be paths in close proximity to the plants, if these are made with gravel or other materials involving lime or limestone. Even leaf mulches from leaves collected in a limestone or chalky area can also affect the colour. Also, simply moving plants from one part of the garden to another into exactly the same soil seems to lose their best blue for a time, for no obvious reason: the disturbance alone seems enough. I have repeatedly noticed blues improving as plants settle after moving. Colour, then, is a sensitive element, meaning that individual experience in individual gardens will vary: so making adjustments to suit individual requirements is essential. Blueing hydrangeas will involve a degree of trial and error.

If a rather muddy halfway-house mauve colour occurs, and you want to intensify it to a strong red or pink, then topdressing with chalk or lime chips, or incorporating these into the soil at planting, will do the trick. Some of the best reds can be as satisfyingly rich a colour as a c ear blue. The old cultivar **'Westfalen'** is one such, and the resulting bright-red colour from such a surface dressing achieves striking effects. But, again, this should not be overdone, as it might induce chlorosis, a yellowing of the leaves due to a lack of some trace elements locked up by the lime. Some red cultivars seem susceptible to this.

Siting and shade

The significant impact of different growing situations on hydrangea health and flowering capacity brings up the prime importance for the entire genus of siting. As a generality, some shade during the hottest part of the day and reasonably moist soil during the growing season are key requirements for optimum performance, for the Macrophyllae in particular. But the amount of shade, in both degree and duration, will affect the flowering performance of all the species and cultivars, one way or another.

Shade is important principally for its cooling effect on air temperature. So the degree of shade that is advisable partly depends on geography: a coastal environment in Cornwall or in Northumbria, for example, where sun is often filtered by a thin veil of light cloud or where a cooling sea fret rolls in from time to time, will moderate high summer temperatures. This means that plants for which shade is an important requirement in inland gardens, often thrive in full sun in such maritime locations.

The vital role played by shade during the recent testing summer in Europe was evident, with temperatures at the leaf surface in the mid to high thirties day after day and a severe want of moisture at the root. Under this pressure, the difference between plants grown in sun or shade was graphically visible.

Two identical plants – one in in dappled shade throughout the day remained in good shape, with both flowers and foliage relatively unaffected (left), while the other, in full sun, was burned and frazzled, even with irrigation (right).

The beneficial impact of shade during excessive summer heat: *H. serrata* 'Tiara' at Wisley. The black line shows the 'tide mark' where the shade reaches.

Birches kept out of the plant beds and confined to the lawn as specimens to avoid root competition still provide a degree of shade.

At the RHS garden Wisley in Surrey, a hedge of *H. serrata* 'Tiara' flanks either side of a short path to offices in a former gardener's house, with a huge old oak to the south providing shade for part of this serrata hedge for much of the day. A 'tide mark' showing the limit of the oak's shade protection is readily visible, where the direct heat from unbroken sun exposure had burned away both foliage and flower.

These observations all hint at the value of preplanning your hydrangea planting places. You need to know exactly where and when your shade falls. If you have trees in the garden it is useful in midsummer to mark the shade limit in the middle of the day with short canes, so you can see where to plant secure in the knowledge that *in extremis* your hydrangeas will have some shade protection when they most need it, at midday in midsummer. The north or north-west wall of a house or outbuilding is another favourable shade situation where hydrangeas will thrive, while also 'anchoring' the house with their mass, and associating nicely, for example, with climbers like roses, whose bony bare legs hydrangeas are good at hiding with neat domes of leafage and flower, while at the same time providing flowers for many summer weeks after the roses are over.

While the value of trees or large shrubs in providing shade is beyond dispute, care has to be taken to ensure that root competition, which dries out the soil in summer, is minimised. Protective shade companions that are surface rooting, like birch or ash, or in a small garden, nearby large shrubs, have this disadvantage. These compete for moisture and nutrients, and can have an adverse impact on other nearby smaller plants.

Magnolias would seem at first to be the ideal shade tree, flowering in spring and offering some top-cover shelter from spring frost, as well as later spreading fraternal branches for welcome summer shade. But they are fierce competitors at the root – their tangled mass of white roots spreading widely and sucking out moisture that can inhibit growth on companion plants, especially those that enjoy more than average moisture. I successfully underplant magnolias with hydrangeas, as well as camellias and rhododendrons, only by locating the edge of the magnolia root system with a border fork, and then as the magnolias grow, root pruning every other year with a sharp spade to hold the line against encroachment in the directions I want to limit. Birches and ash, however, are such brutally competitive surface rooters that all hydrangeas should be kept well clear of them, even with good irrigation.

If there are no large trees to provide shade in a small garden, upright and spreading shrubs, such as forms of *Acer palmatum*, the Japanese maple, or the more vigorous forms of *Philadelphus*, can be cleaned up or 'lifted' to soon form clean-trunked, multi-stemmed, small trees as local shade and shelter for the benefit of *H. macrophylla* or *H. serrata*. Some shrubs look attractive when lifted in this way and at the same time supply necessary space and air for many kinds of companion plantings, both under and near. For contrasting masses of colour, vigorous cultivars of *H. paniculata* can themselves be so 'lifted', enabling white/pink flowers to enhance the brightly coloured forms of the Macrophyllae subsection planted beneath.

Some hydrangeas will flower generously even in quite dense shade. These shade-bearers will flower freely in poor light, even if never seeing direct sunlight at any time of day. They are thus of exceptional value in deep woodland, where the Macrophyllae are likely to be drawn up out of character and to

A *H. paniculata* pruned to a small tree to allow companion plantings to benefit from topcover.

The 2008 RHS *H. paniculata* trial was in full sun and exposure, with flower on every shoot.

flower only sparsely. Most notable and quite exceptional as shade-bearers in this respect are the cultivars of *H. involucrata*; but *H. aspera* and its varieties can also enjoy partial shade, although typically flowering more freely with good light and an open situation. Conversely, *H. paniculata* and *H. arborescens* are not such willing shade-bearers, often producing blind shoots in shade (with no flowerbuds); they will perform in partial shade with full light overhead, but to flower freely will always prefer plenty of sun. These later two subsections of the genus are the tough heat- and cold-resistant species that not only will take full sun in a hot summer without flinching, but in an average summer will flower incredibly freely in the open with flower on every shoot.

Practical growing tips

Soils

Most hydrangeas are very tolerant of most soils, and even in extreme soils – heavy clay, light sand and chalk – with care, hydrangeas can flourish. But like any woody plant they will not put up with wet feet during their dormant season, especially in a wet winter. This is a well-known truism but it bears repeating. The compact, thickly matted root system of all hydrangeas demands decent drainage to ensure a good supply of oxygen throughout the rootball: anaerobic conditions over weeks will quickly rot hydrangea roots. This means, for example, that highly retentive heavy clay soils liable to ponding need the benefit of additional materials that will keep the soil open and breathable, and allow the release of surplus water: generous inorganic additions, such as grit or coarse sand, are essential. My late father used to wrestle with sticky Northamptonshire clay, which, although growing excellent wheat and top class beef, made his gardening life difficult. He used to make his own terracotta from it, as a garden soil additive: he would pack clay sods dug from his field over a hot log fire, leaving it to heat and bake for days, eventually producing a crumbly inert substance like a soft and light pulverised brick. This would never return to its clay consistency and, dug in in quantity, it was highly effective in helping keep his clay soil open.

Organic material such as garden compost, well-rotted bark and woodchips, peat or peat substitutes and leaf-mould will also keep the soil consistency open and add nutrient for good growth. Care must be taken to avoid creating a sump underneath the layer of prepared soil: in clay it is good practice to aerate the subsoil by turning it over thoroughly with a fork, and add to it a layer of grit before backfilling with organic material such as compost and leaf mould mixed in, to ensure good drainage of the planting hole.

Apart from clay, the other challenging soils for hydrangeas are light, sandy soils with poor water-retention that dry out rapidly in summer, and soils on chalk. The objective here is the opposite to clay – to hold water – but the remedy is the same: generous amounts of organic material, such as leafmould or compost, are essential to help retain moisture at the root; and in the case of chalk soils, to help remedy any lime-induced chlorosis. This can also be mitigated by watering with a proprietory iron chelate, available in most garden centres. Mulching at the time of planting and regularly thereafter is a vital remedy to help to achieve a good result from such extreme or challenging soils.

The ideal soil is a moisture-retaining but well-drained medium loam, containing plenty of organic material, preferably with a pH below 6.5 to allow blue flowers, as discussed above. Contrary to popular belief, this may be the most frequently encountered soil type in the UK.

Ideal planting hole preparation: (a) digging down to subsoil; (b) loosening subsoil with a fork; (c) mixture of leaf mould and garden compost added to the backfill; (d) pot-bound roots; (e) loosening root system with a handfork; (f) firmed with fingers, not traditional gardener's heel; (g) adding mulch.

Mulching

The only garden by-products we burn at White House Farm are unrottable tree roots, rose prunings (for disease prevention) and *Berberis* clippings (for potential punctures, both mechanical and human). Everything else is used, the vast bulk chipped or composted for re-use as planting material or mulch. Mulching is now proven beyond argument as our most valuable tool for longer term soil improvement and good growth through soil structure betterment and enrichment. It feeds the soil both physically and chemically. Its short-term benefits are obvious – weed suppression and moisture retention through evaporation prevention.

Mulching is particularly valuable for newly planted, surface-rooting hydrangeas that require plenty of moisture at the root during their growing season. We habitually pile about 12cm (5in) of mixed, partially broken-down woodchip, leaves and bark around hydrangea plants, across an area about 1m (3ft) in diameter, avoiding burying the stems (stems need a generous 18cm/7in clearance on first application). Ideally, we mulch with bark and woodchips and chipped prunings: once applied, this thickness normally lasts for a year. We try to prune trees from July onwards: one of the benefits of pruning them in leaf is the addition of foliage to the woodchip, both contributing to rot down and enhancing nutrients. It is a crime to burn gathered autumn leaves, which can be swept up and directly added to plants, to rot down and be pulled in by worms *in situ*.

A side benefit of summer pruning is that chipped branches include leaves with the wood and bark, making an ideal mulch.

And grass cuttings do not need composting either. If applied directly round the plants in thin layers of no more than 6cm (2½in), they dry and quickly turn to an anonymous brown, avoiding the sticky mess accruing from piling them thickly or in heaps. They are excellent annual weed suppressors and keep moisture in very effectively during warm spells. When using all these materials as a mulch, none need be fully rotted; indeed, we have used them all fresh with no obvious deleterious effect. But it is, of course, a given that they remain a surface cover. Mulch, by definition, is material that is never dug in.

Annual weeds will germinate in mulch sporadically, but are root insecure and thus easily removed. With the increasing prevalence of hot, dry summers in the UK, the value of mulch in preventing evaporation, reducing watering and maintaining plants in good health will be ever more important. Some commercial mulches on offer can have a high (alkaline) pH, so if you want blue hydrangeas, it is worth checking the pH with the supplier before placing a bulk order. The same goes for leaves collected from limestone or chalk locations.

Slope and air drainage

At the same time as planning siting, soils and mulching, consider slope. Dense, cold air is heavier than that of normal temperature, so good air drainage is highly desirable, allowing the cold, heavy air of late radiation frosts to flow away downhill. This is a key factor to avoid damage in spring and autumn when a light late-spring or early autumn frost, if plants are actively growing, can do more damage in one night than an entire cold winter. This tendency of cold air to sink is known as the katabatic wind (from the Greek '*kata*' meaning down and '*bainein*', to go or walk) and will roll away downhill if given the opportunity of an unencumbered escape channel.

I once experienced a dramatic example of katabatics on a large scale, camping with the children in August on the northern edge of Lake Maggiore in north Italy: as dawn broke every morning, a brisk cool wind would arise as the dense cold night air from the high slopes of St Moritz, about 112km (70 miles) up the mountain to the north, came rushing down, ruffling the lake and enticing the fishermen out in their little boats. As the sun emerged during the morning and began to warm the environment, a stillness. Then in the afternoon the breeze would reverse direction, as the air to the south on the extensive Milan plain was warmed by the day's summer sun, and the warm air rose and flowed gently up the valley.

So installing hydrangeas at the top or the side of a slope is good practice, if such is available. Frost pockets at the bottom of valleys are to be avoided at all costs, tempting though it might be to seek the benefits of shelter from wind and good soil moisture. It is worth bearing in mind that hydrangeas are relatively wind-tolerant, compared to say, roses or rhododendrons.

Late-spring frost can inflict more damage than low winter temperatures.

The katabatic effect can apply on a very small scale in the garden, where even quite small undulations can have a similar influence. After a sharp, late-spring frost, very often a 'tide mark' of frost effect can be seen on plants, marking the settling and ponding limit of frosty cold air on fresh foliage.

Planting on a slope requires the construction of a level or flat area for the planting hole itself, to retain moisture and hold rainfall, so some modest terracing is required for each plant. This is essential, as failure to do it will either lead to drying out in summer as rain runs off down the slope without penetrating, or the plant's stem being overwhelmed by soil slipping down to bury the collar and induce decay. Always plant on the level, with, if anything, a slight camber towards the back of the planting hole to hold in the rain.

Tree or shrub topcover is doubly protective, helpful against both spring frost and providing partial shade in summer, and creating the humidity that the plants enjoy. Shady wall protection can serve a similar function, as well as shielding from wind. Freezing winds are the most damaging factor in hydrangea growing. I have seen plants that withstood a frost of −10°C in still conditions without a problem, twig-killed down to two years' growth level by −5°C minimum a few years later by an exceptionally searing easterly wind with frost on its breath that blew for days. Wind shelter from north and east is paramount.

Planting on a slope with terracing.

Pruning

Most advice on pruning, even today, is focused on *H. macrophylla*, apart from some encouragement to butcher poor 'Annabelle' and undertake a major operation on *H. paniculata*. Much advice on hydrangea pruning can sound complicated and be rather offputting: the first thing to say is that it is not necessary to prune at all for a good few years after planting. Many *H. macrophylla* plants will require no pruning at all – I have a 45-year-old plant of *H. macrophylla* 'Mme Emile Mouillere' still growing freely in part-shade and the only pruning has been to remove the front of the spreading plant with a hedge trimmer to avoid its invading and eating up more lawn.

The plant itself will tell you when to prune, as its lower branches become overgrown, overshaded and die out, just like a shrub rose. This dead wood should, of course, be removed in its entirety. Over the passage of time, maybe five or six years, all fresh *H. macrophylla* shoots will eventually mature, hardening, thickening, developing rough bark and losing their vigour as senescence sets in. These ageing shoots can also be removed in their entirety from the base with loppers, leaving younger vigorous shoots more space and energy to grow-on and recreate the bush. As someone said of a garden rose, it is only as old as its oldest wood. This is also true of a hydrangea. They are long-lived plants, when happy and kept young.

Deadheading

Some authorities suggest that spent dead flowers, especially conspicuous on the globose hortensias, should be removed in autumn both to improve the look of the bush and also to allow the shoots bearing flower-buds for the next year to ripen off and harden into maturity in the sun and wind before winter arrives. This may be a good procedure in softer climatic zones where frost is less of a problem and growth is more lush, but others, myself included, prefer to leave them on the bush over winter in the belief that they offer some protection against potential frost damage to the fresh shoots growing out immediately below. They can be removed for aesthetic reasons as soon as the danger of spring frosts has gone. Some gardeners find them quite attractive over winter and it is certainly true, decorated with a heavy hoar frost, they can sparkle prettily in the low sun.

With blue *H. macrophylla*, especially where they are difficult 'bluers', it is also good practice to push the removed dead heads well under the bush as a mulch and allow them to gently rot down *in situ*. They contain the aluminium that creates the blue colour and it is important to recycle it, not take it away from the plant.

If *H. macrophylla* or *serrata* plants are seriously damaged by extreme heat or cold, or simply get too congested over time by a thicket of growth, they can be 'started again', by cutting down to about 30cm (1ft) high in late winter. They will shoot freely to create a new bush as spring temperatures warm – often too freely, and it will be necessary to thin the new shoots to provide enough air and sun through the centre of the plant to ensure adequate wood ripening before next winter sets in. A more open or airy bush structure, which allows all shoots to harden up in their first season, is essential in a colder garden. This 'starting again' means losing almost

Hydrangea macrophylla 'Mme Emile Mouillere' in September.

all flowers for a season, but in the following year it will pay dividends with an exceptional display, as well as rejuvenating the whole plant. In broad terms these suggestions also apply to the Asperae (Chapter 5).

In the section on *H. paniculata* in Chapter 4, I discussed how the timing and degree of pruning can affect the timing and size of the flowers. This was one of the useful supplementary bits of information to come out of the RHS 2008 *H. paniculata* AGM trial. The harder the pruning, the bigger and later the flowers; and the later the pruning, the later the flowering time: so a hard, late pruning can delay flowering by some weeks, as well as producing large flowerheads. An early, light prune or simple deadheading will give a notably different result to a late, hard prune, with more, smaller flowers produced early. It is not necessary to always cut the plant hard back, as often advocated, unless you deliberately seek the very largest possible flowerheads. But bear in mind that many of these massive, heavyweight, solid *H. paniculata* heads can weigh a kilo when wet and flop to the ground in bad weather. Pruning large-flowered *H. arborescens* cultivars follows the same logic (Chapter 3). The seminal 'Annabelle', for example, can be varied in size according to gardener taste, determined by the chosen pruning regime.

Dead flowers left on the bush can help protect emerging shoots from untimely frost.

CHAPTER 11

PROPAGATION

Hydrangeas from seed

Growing woody plants from seed is one of the most rewarding experiences in horticulture. Outcomes in flower and foliage quality are usually uncertain from seed, and the pleasure of anticipation is intense as the hour approaches when the first flowers open, usually after several years. The late Peter Chappell used to say, mischievously, that growing flowering shrubs was a question of eleven months' pleasurable anticipation, followed by one month of intense disappointment. The fact that he took great pleasure in growing a wide range of woody plants easily gave the lie to this and, in particular, it is not at all true with a genus as genetically unstable and varied as *Hydrangea*. Most results are worthwhile and some are really good. The excitement of anticipation is further intensified if a successful deliberate cross has been made with the objective of achieving a specific result.

Hydrangea seeds: you can hold 100 plants between thumb and forefinger.

Hydrangeas are something of a challenge for the amateur as the seeds are so tiny, with very few species producing seeds exceeding 1mm long. In terms of human scale, they are, essentially, dust. I can comfortably hold 100+ hydrangeas between my thumb and forefinger.

There are many different ways of dealing with these successfully, each individual gardener having their own preferred method. My own procedure is to use a small home-made propagator in the cellar, whose advantage – as for wine – is the steadiness of the ambient temperature, which is a consistent 10°C (50°F) summer and winter, night and day (thanks to the boiler and hot-water tank in the corner). Those stable temperature conditions often apply to other rooms in the house, for example an empty bedroom or cloakroom.

I built a small propagator about 2 × 1m (6 × 4ft) using angle iron and plywood as the base, adding table-height legs and fitting a 15cm (6in) deep tray secured to the structure, which I filled with sterile sand to make a sand bed, covered with a polythene tent. This was furnished with bottom heat from a heating cable run underneath, and two 1.5m (5ft) strip lights fixed at about 45cm (18in) above the seed trays to provide adequate

OPPOSITE PAGE A vivid dark blue White House Farm seedling.

light and prevent etiolation. The sand bed is kept permanently damp but with sufficient air allowed into the polythene tent to inhibit the growth of moulds and fungi, and notably 'damping-off' (caused by a mixture of several of these). Once damping-off seedling disease gets into a seed tray it can be devastating, rapidly consuming everything in its path, like an unstoppable forest fire. To inhibit it, the humidity should be a little less than 100 per cent, seed trays must be sterilised and the growing medium inert. Above all it is vital not to sow too thickly – which with small seeds like hydrangea is challenging and requires discipline. I then set a 14h day on a timer switch to turn the lights on and off, and control the sand surface temperature thermostatically at a brisk 20°C, plus or minus 2°C, the variation on a simple rod thermostat. A similar professional system may be purchased off the peg for a few hundred pounds. You just have to plug it in.

For the inert medium, I fill shallow plastic trays 5cm (2in) deep to within a centimetre of the top with a 50/50 mix of peat (or peat substitute) and perlite, and this is sterilised in a microwave oven for two and a half minutes. Much longer than this and the plastic trays begin to distort, then melt. It seems to work well: over many years of constant germination using this method I have not seen weed seeds germinating and I have never seen a bug of any description.

The hydrangea seed is sown on the surface of the compost and not covered, but instead watered in with a fine spray to sink the seeds into the surface and ensure that they are well wetted. A rigid transparent plastic sheet cut to size is placed on each seed tray to prevent moisture loss, the gap between compost and plastic lid being about 1cm (½in), so the seeds, once wetted, never dry out. This is turned every couple of days. I use the same system for other very small seeds, such as Deutzia, Hypericum, Rhododendron and Philadelphus.

Seed is usually ripe on most species of hydrangea by December and ready for collection. The small capsules are normally ripe and brown, but not yet open. Collecting and sowing procedures will vary with individual preferences and the rule

Plastic trays with inert compost ready for sowing, with the customised rigid plastic cover.

Using a sheet of white paper to collect hydrangea seed. A single lacecap inflorescence can produce thousands of seeds.

is – if it works, stay with it. For what it is worth, I take a whole inflorescence and place it upside down on a sheet of matt white paper in a dry, but relatively cool place, such as a spare bedroom – somewhere that has the advantage of not being vulnerable to passing human traffic. A big sneeze, for example, or a swishing coat could wreck in an instant all the earlier painstaking work undertaken in making a cross. After two weeks, the capsules will open and seeds will be visible on the white paper. A gentle tap of the inflorescence will release more, usually much more than is needed, especially with lacecaps. This is a particularly useful method for mophead *H. macrophylla* cultivars, whose fertile flowers are sparse, hidden and whose seed quantities much smaller: but they show up well on the white paper.

I sow the seed between Christmas and New Year, and germination normally takes place in about 11–14 days. I believe this method saves a lot of time, as by April, when you would be expecting initial germination under normal outdoor temperatures, I am already pricking out seedlings into small liner pots, and weaning them for a few days under a very thin polythene sheet in a greenhouse. Usually, depending on the vigour of the taxon, I would expect to be potting on again into 1ltr pots by the end of July or into August. Having overwintered in the greenhouse for their first winter, young plants can then be transplanted to nursery beds in a favoured part-shaded situation outside to grow-on during the following year after the potential danger from spring frost is over. One or two may flower in the following third year, others in year four. Then the tricky job begins of deciding if any of them have merit and are worth propagating. I tend to favour a wait of about ten years for a full trial, though selections have been made sooner.

It is well worth taking the trouble to grow any hydrangea from open-pollinated seed, without the fiddle and fuss of hand pollination (for more, *see* 'Hybridising' later in this chapter). The variety of form and colour will almost guarantee interesting, if not earth-shattering, results; many of the Japanese-named varieties of both *H. macrophylla* and *H. serrata*, for example, are often naturally occurring hybrids or anomalous forms, and some of the results from seed can be surprisingly innovative and attractive. *Hydrangea aspera* can yield surprisingly good results in both flower and foliage, as well as hardiness. It is a most satisfying and rewarding part of horticulture and I warmly recommend it.

Hydrangea cuttings

Hydrangeas are among the easiest of woody plants to grow from cuttings, especially the colourful and popular cultivars of *H. macrophylla* and *H. serrata*, which are the easiest of the easy to propagate by this method.

Cultivars of all the other hydrangea species will also root readily, but the species with hairy or pubescent leaves, like *H. aspera* and *H. involucrata*, may need a little more care in timing and aftercare management to prevent damping-off and successfully bring them through winter dormancy after rooting. The dormancy period can be tricky for the Asperae, in particular, as *H. aspera* rooted cuttings, ripening later in the season, have a shorter window to put on enough growth to build up sufficient carbohydrate to see them through the winter months, and sometimes simply gently wither away during dormancy. The best solution is to keep them dry, in a cold but frost-free environment, and not to pot them on until growth resumes in March/April. Alternatively, pre-empt dormancy and keep them growing under artificial lights, tricking them that it is still autumn, until well into December.

Hardwood cuttings taken from fully ripened wood can root. Indeed, almost any piece of hydrangea simply stuck into the ground in winter, especially in sandy soil, may root, but it is not a fully reliable method and results can be patchy and growing-on rather slow. The best and easiest method to propagate hydrangeas is with summer cuttings, taking new growth that is just beginning to ripen and become firm. These will succeed when placed either in a mist system or simply under polythene in a closed case. Younger cuttings, still soft and sappy, can succeed under mist, but are more likely to damp-off in the more closed, static 100 per cent humidity of a closed polythene propagator. For the amateur, normally just using a simple arrangement with a polythene tent of some kind, cuttings taken from May through to September can succeed, though the earlier the better, to give the plants more time to grow-on and establish well before seasonal daylight length and a lower temperature bring on dormancy. A simple pot covered with a polythene bag secured with an elastic band can work for those with no other available space.

Hydrangea serrata cuttings rooting in moisture-saturated air in a warm closed case, demonstrating how easy they are to root.

Minimal material can root!

I had a big plant of *H. macrophylla* 'Mariesii' that produced a single beautifully variegated leaf with yellow edges and streaks. I wanted to propagate this interesting foliage sport, but it was only a single leaf! I cut just above and below the leaf and its axillary bud, leaving about an inch of stem below, and inserted the single leaf-bud as a cutting, burying it on its side at an angle so that the bud was at compost level. It rooted nicely, grew out as hoped as a variegated shoot and grew on well, but eventually proved not to be stable, irritatingly reverting to produce green as well as variegated shoots. I removed the reversions and rerooted a couple of the variegated shoots but the instability persisted, and in the end I gave it up as a bad job and my striking new plant of *H. macrophylla* 'Mariesii variegata' never materialised.

The state of the wood of the cutting is the key to success, rather than the calendar, though hydrangeas are particularly forgiving and I have successfully taken cuttings as late as October. I recall the late Maurice Mason, when I mildly objected that it was too late when he offered me October cuttings, saying, 'Strike while the iron is hot; always accept an offer, you've nothing to lose'. He was right and the cuttings rooted and grew on.

A standard cutting should ideally consist of a shoot with three pairs of nodes in the axils of the leaves. The leaves should be removed from the lowest pair of nodes and the cutting severed immediately below, where there is a natural concentration of hormones. The remaining leaf tips should be clipped to reduced their surface area by a third to a half. This facilitates handling and makes it easier to keep the cutting upright in the compost after insertion. A dip in a proprietory hormone rooting powder may be useful, though with easy rooters like hydrangeas is by no means essential. A thin sliver of bark should be removed about 3cm (1in) long on one side of the base of the cutting and it should be inserted shallowly, leaving a small 'church window' of removed bark visible at the surface. This will show when the cutting has calloused and is a useful signal to impatient propagators as to progress.

Preparing cuttings: a thin sliver of bark can be left protruding from the soil to monitor callousing.

Hydrangea cuttings after three months.

If material is scarce, say from a single donated shoot, short internodal cuttings will succeed; that is, with the cutting severed between nodes rather than a shoot severed at a pair of nodes. One pair of leaves will suffice to assist rooting. This will maximise return on minimal material.

Cuttings should be ready to pot on in 4–5 weeks. It is preferable to err on the generous side regarding time before

Two rooted internodal cuttings cut from a single donated shoot, showing rooting after five weeks under polythene.

Standard cuttings ready for potting on after four weeks.

Grafting hydrangeas

Grafting is only very rarely used, if at all, and only if necessary, as cuttings are so cooperative in rooting well and hydrangea bark is thin. A skilled propagator friend sent me a rare double form of *H. heteromalla* that he had no choice but to graft, as material was sent to him during winter dormancy and he wanted to be sure of a successful propagation. In fact it did not grow-on very well, so I soon grazed it for cuttings, and these made strong plants. But it did demonstrate convincingly that grafting is possible.

transplanting, leaving cuttings in the pot for a week longer if in any doubt, as a strongly rooted cutting will establish more effectively and grow-on more strongly when potted up.

It is optional whether to leave or remove the terminal bud: in my experience this seems to have no effect on speed or ability to root. If left on, it will be the first bud to expand and can help indicate that rooting is sufficiently advanced to pot on. And after rooting it will extend and grow-on quite quickly when transplanted.

Bamboo-shoot pegs for layering.

Layering hydrangeas

If only a few plants are required, then layering is another very easy mode of increase. In fact, old plants can self-layer over time, especially the more vigorous sorts in good conditions, with leaf fall becoming trapped inside the plant and providing a covering layer. I have a single 40-year-old plant of *H. macrophylla* 'Mme Emile Mouillere' now covering an area some 4m^2 (42ft^2), gently advancing with self-layered branches.

The prescribed method of layering is to select a suitable branch, scuff the bark at the junction of new and old wood, make a shallow trench in the appropriate place, place peat or moisture-retentive compost in the depression, peg down the layer firmly, cover with soil, stake the shoot and tie it in vertically. If kept moist, layers made in the spring should have rooted sufficiently to be severed the following season. If you have access to nearby bamboo, an excellent peg can be contrived from taking a length of a bamboo shoot and cutting it below a joint and about 15cm (6in) above the joint. Clip off each side shoot to around 8cm (3in) and simply turn it upside down. *Voilà!*

For *H. macrophylla* and *H. serrata*, which root from layers very readily, a simple method using two old house bricks, one placed over the branch and the other pushed against it trapping the shooting end of the branch in the gap to hold it up vertically, will do the job very well, with the bricks effectively holding in moisture. But a sharp eye needs to be kept out for slugs, snails, earwigs and woodlice, as they are attracted by the cool, damp bricks and can wreak havoc quite quickly.

Hybridising hydrangeas

On a warm July day, a planting of *H. aspera* forms is abuzz with insect life of all kinds, feasting on pollen and nectar abundantly available on the wide, flat, easily accessible discs of fertile flowers. As their *quid pro quo*, hover flies, honey bees, bumble bees, butterflies, moths and myriad small unidentified black beetles and creepy crawlies all do their work of pollinating the plants. This means that many seedlings raised from open-pollinated seed collected from your favourite cultivars may be hybrids, with the insects possibly bringing pollen from neighbouring plants. In fact they may be hybrids twice over, many having been hybrids in the first place, so thoroughly mixed up and mongrelised. Simply taking seed from hydrangeas grown in varied outdoor contexts gives good chances of interesting, often entirely unexpected results – even to the extent of producing something new in form or colour, or both.

I have sowed open-pollinated seed from a variety of Japanese forms of *H. serrata* with exciting results, different in colour and form from their parent plant, with some certainly worth naming and propagating. A small number show the obvious influence of *H. macrophylla*, with the heavier foliage, taller habit and stouter shoots of that species, a hybrid contamination either belonging to the original plant or as a result of concomitant insect activity in the garden (*see* Chapter 13). Other *H. serrata* seedlings from open-pollination are intermediate; still more show the characters of the *H. serrata* parent. The point is that sowing seed is easy, fun and can actually produce worthwhile results.

To produce a specific and planned result through controlled hand-pollination, I select two appropriate parents. I normally reduce the whole corymb of the female parent to some five still unopened, fertile flowers, removing the rest of the corymb with a pair of sharp nail scissors, then marking and labelling the cross. As the flowers begin to open, and before the stamens produce pollen, it is vital to prevent self-pollination by removing the stamens with scissors. At the same time, I use a suitable porous bag to protect the maturing flowers from pollen contamination brought by the myriad insects of summer; this has to

Breeding *H. aspera* for translucent foliage colour.

be removed to inspect the stigma to determine when it is receptive. Pollen from the male parent can then be dusted on it. I normally simply apply pollen direct from the male flower without using any intermediary such as a light brush. I usually leave the pollinating flower elements in the protective bag for some days as a belt-and-braces gambit. I then wait, hoping to see a capsule developing, which indicates that pollination has actually taken place and seeds are forming. Months pass until December, when the seed can be harvested. It is a fiddly job, but highly satisfying when successful.

Simply sowing open-pollinated seed certainly avoids the rather fussy and fiddly procedures of deliberate hand-pollinating. But it is even more satisfying to select two parents, cross them with the deliberate intention of achieving a specific objective, then securing a happy combination of the best qualities of both in the ensuing seedling.

For example, I hand-pollinated a cultivar of *H. aspera*, grown from seed collected in the wild, which had a richly coloured red/purple reverse to the leaf and was arborescent, but not reliably hardy. It was killed back to three-year-old wood by an exceptional searing easterly wind. The plan was to cross it with a hardy plant that would preserve its foliage colour and its tendency to make a small tree, while at the same time introducing improved hardiness. The idea was to develop a standard form of *H. aspera* as a tree with coloured leaves so you could look up into the foliage and see the light shining through the leaves from above, to show off these enchanting colours of orange and bronze, as through a magic lantern.

The result was mixed in terms of the objectives, and not entirely satisfactory. A few plants kept the arborescent character, making single stems in the nursery bed with little tendency to throw basal shoots; but those that did keep the treelike habit all had a paler leaf reverse than either parent. Hardiness has not yet been put fully to the test, as the ensuing winters have been mild. The bushier non-tree-like seedlings resembled either parent in their dark-reddish leaf underside and were perfectly good plants, but they were not a step forward from the parent. I have kept three 'trees' now coming into year five, pruning them as standards. So far so good: time is always a key dimension in the hybridising game.

Macrophylla influence in a *macrophylla* × *serrata* seedling.

CHAPTER 12

PESTS AND DISEASES

I have been growing hydrangeas for over 50 years and have never had occasion to spray them against either bugs or bacteria. They are among the most trouble-free of all garden plants. Vigilant observation is the best answer.

Pests

Even rabbits prone to consume most sorts of vegetable growth quite indiscriminately, hardly ever touch hydrangeas. Just occasionally a new shoot can be seen lying next to its parent, unconsumed but with the trademark clean chiselling at 45 degrees that marks a rabbit with time on its paws. In this they are looking not for food, but simply for tooth exercise.

Even deer prone to prune any kind of rose to the ground and rub valuable young saplings to extinction with their antlers, generally eschew hydrangeas, perhaps instinctively detecting a noxious presence of cyanide, which some botanists associate with the genus: not at all good for Bambi. I have found that bamboo canes, with rag strips bound and knotted round the top, soaked in some smelly water-miscible phenol, like Jeyes Fluid or Vinyl, and stuck in the ground to blend into the plant is something of a deterrent, for both rabbits and deer. After a time the smell is not detected by passing humans.

Moles can be a nuisance, as they are always attracted by specially prepared soil for a newly installed plant, and by damp soil after irrigation. They can loosen the soil around the roots and occasionally 'heave' a small plant. The answer is to trap them using either a barrel trap in the deeper runs or the easier scissors' trap set in a surface run near a molehill. There is usually a local specialist who can undertake this. I know of old hands at this game, seething with frustration, who wait in the early morning with a shotgun and blast the moving heap as the mole hits the surface. Some succeed by driving a spade into the moving heap and throwing the soil in the air, complete with mole.

A teeming population of red ants can destroy the root systems of newly planted hydrangeas, especially during a warm spell in a light, friable soil, and plants can die before the disturbance is noticed. Any small plant beginning to flag should catch your attention, and its surface soil examined to see if it has been built into a fine sand-like heap around the base of the plant. I lost a valuable one-year graft of the rare *Berberis montana* to ants, failing to notice the massive disturbance of root activity in time. The application of a propriety ant powder dusted around the plant will solve the problem. Ants can also spread aphids, 'farming' them for their sugary exudations and introducing them into new pastures.

The usual pesky small fry can do minor damage: capsid bugs can deform shoot tips and burgeoning leaves and inhibit flowering, as well as creating an unsightly look, but this is not

OPPOSITE PAGE: 'Hirose No Hana', a Japanese cultivar of *H. serrata*, with long-lasting flowerheads packed with mauve/pink and cream star-like petaloid processes.

common; thrips can also put in an occasional appearance. Each and both of these bugs could be the one occasion to spray some controlling systemic remedy. Aphids are only occasionally seen and the finger and thumb method provides a good solution. Ladybird larvae help but usually do not finish the job if there is a big infestation. Biological controls can be very useful under glass, or even the old remedy of sticky sheets hung between the plants.

The significant destructive vandals for hydrangeas are slugs and snails. They can make short work of a small plant overnight. They have a particular appetite for *H. macrophylla* and *H. serrata*. I have also lost plants of *H. heteromalla*, where large slugs scoured off all the soft basal bark and effectively cut the food and water supply. There are many remedies proposed, from eggshells to copper bands to grit and pine needles. None of them work and with a large collection, none are practicable. The most convincing answer I have found has been a concoction of garlic, where garlic cloves are mashed, soaked in water, the resulting liquor diluted, filtered and watered on and around the plant. It seems to work even on vulnerable and tasty hostas. The other effective option is to use slug pellets, though care has to be taken to use only the kinds that do not damage pets, song thrushes (the only garden species to eat snails) or hedgehogs.

Scale insects can be a problem, though not a common one. There are many species, some of which exude a sticky sugar-rich honeydew that encourages the growth of sooty mould, which can be highly disfiguring and weaken the plant's ability to photosynthesise. Both scale and sooty mould might be found on hydrangeas grown under tree species that are particularly susceptible to scale, such as sycamores and limes.

Newer pests are evidently emerging in the form of alien beetles of various kinds, although in fifty years of growing hydrangeas outdoors in Kent, I have never seen evidence of their presence. The accompanying photographs, showing extensive damage to the leaf and ray-flower edges, were taken of a newly planted mixed bed in a central London park (SE1).

Pulvinaria hydrangeae – a scale insect particular to hydrangeas.

A newly planted mixed bed in a central London redevelopment shows that beetles are attracted to both hydrangea flowers and foliage.

Diseases

Most hydrangea diseases are usually a function of a fault in cultivation, and not normally encountered when plants are growing well in the garden. For example, dryness at the root, too much shade, too much sun, poor drainage or planting too deep: any of these conditions can weaken a plant and make it vulnerable to fungal attacks of powdery mildew, root rot or botrytis. Unless the problem is severe, which is unlikely for plants outdoors, it can be remedied by common-sense improvements in cultivation like watering, mulching or, in the case of root rot due to poor drainage, more drastic treatment: transplanting. With care to lift the plant together with a good rootball, and proper post-transplant precautions like canopy reduction, watering and temporary shade, this can be done at any season, even with the plant in full leaf. Plants in the greenhouse or tunnel are generally more susceptible to fungal diseases, perhaps because of their over-fertilised and more humid and warm atmosphere – conditions conducive to fungal growth. The best answer here is to spray with a proprietry fungicide.

As the summer wears on, the cumulative effects of heavy rain, hail, strong winds, drought and heat have an inevitable impact on the appearance of the plant and can be increasingly in evidence, particularly on a hydrangea in exposure. It can be easy to mistake the occasional summer weather disfigurements of foliage in the rigours of the open garden for some kind of disease. The browning of old leaf edges, the partial dessication of fresh young foliage, some minor twig dieback can look like evidence of an endemic problem when it is just a question of wear and tear, simply requiring a little tidying up.

Fastidious, uncooperative, hard-to-please *H. macrophylla* 'Merveille Sanguine': not diseased but often needs a wash and brush-up.

Mechanical damage: a hydrangea aspera leaf after a heavy hail shower.

Cercospora leaf spot is a recently reported fungal pathogen from North America, where it evidently favours more southern locations. It is not a chronic problem at White House Farm, where the only cultivar to be affected appears to be the old Japanese hybrid *macrophylla* × *serrata* called 'Warabe'. I have not seen it here on arborescens, paniculata, or macrophylla – indeed, the old macrophylla cultivar 'Harry's Red' grows through and into afflicted 'Warabe' with complete impunity, showing no inclination to develop the disease. Its spread can be checked with a suitable fungicide but, in UK conditions at present, it does not appear to be a major threat.

'Warabe' is the only cultivar at present to be seen with the *Cercospora* disease in our representative collection at White House Farm.

CHAPTER 13

FUTURE OPPORTUNITIES FOR DEVELOPMENT

The development of new hortensias and paniculatas may be reaching the point of saturation. With these species, it is only the flower, in the end, that sells the plant. While many fine varieties have been introduced in the last decades, and 'novelties' still continue to flood on to the market, the range of variation in habit, flower colour, size and form within the confines of *H. macrophylla* has been almost infinitely exploited; and most of these newly marketed plants, some of which are developed specifically for the florist's

Hydrangea macrophylla 'Miss Saori', plant of the year at Chelsea 2014.

OPPOSITE PAGE: Exceptionally free-flowering, with purple, young growth, *Schizophragma hydrangeoides* (now *Hydrangea hydrageoides*) 'Snow Sensation' arose spontaneously from a sport (mutation) from a pink form ('Rose Sensation') in production at the Minier Nursery in France. © Hortival Diffusion

There are many colourful and promising new introductions still to be tested fully in garden conditions. (a) 'Baron Pourpre'; (b) 'Can Can'; (c) 'Curly Wurly'; (d) 'Elegance'; (e) 'Felina Blue'; (f) 'Frou Frou'; (g) 'Red Ace'; (h) 'Royal Red'.

shop, still remain to be fully tested in the grate and grind of the open garden, through wind, sun, rain and the rigours of winter.

For example, the beautiful *H. macrophylla* 'Miss Saori', which won plant of the year at Chelsea in 2014, has turned out to be rather less than weatherproof in UK garden conditions.

Like many market-driven commodities, they may strut their stuff for a while, enjoying the brief limelight at front of stage, only to be gradually forgotten and discarded in the remorseless test of time. Like fashion in any other walk of life, such as pop music, they are here today and gone tomorrow, with only a few staying the course for the long run. It's the survival of the fittest – the fittest for long-term garden ornament, for the gardener looking to permanently enrich his or her summer garden.

Some hydrangeas have been with us for well over 100 years, stood the test of time and are still in demand. (a) 'Mme E. Mouillere'; (b) 'Bluewave'; (c) 'Foch'; (d) 'King George'; (e) 'Générale Vicomtesse de Vibraye'; (f) 'Mariesii Grandiflora'; (g) 'Mariesii'.

The breeding potential of *Hydrangea serrata*

Historically, some hybrids already exist between the closely related *H. macrophylla* and *H. serrata*. These are not only a significant innovation, but there are likely to be more to come, with deliberate hybrids potentially able to exploit the floral opulence of the soft coastal hortensia with the more subtle variations of the cold-tolerant montane species. *Hydrangea serrata*, in general, has yet to be fully appreciated outside Japan, though it is now beginning to make serious progress among discerning European gardeners. Many professional commercial growers still have not fully grasped that, as an essentially montane species, it is among the hardiest in the genus, surviving winter lows that belie its ostensibly fragile, thin-twigged, small-leaved persona and allow it to succeed where the maritime *H. macrophylla* will fail. It also has a shorter growing season than the long growth span and thus frost-vulnerable 'big leaf' (macrophylla) hydrangea, starting later and finishing earlier, entering winter with properly ripened wood. It is as tough as an old boot.

My own amateur attempts at hybridisation, growing-on for trial only scores rather than the hundreds of seedlings usually set out for selection by institutions and commercial hybridists, have yielded quite a few plants of merit, given very good marks by visitors. Simple open-pollinated seedlings from *H. serrata* have also produced some excellent flowers of varied colour and form to rival some of the Japanese selections that have so far reached Europe (*see* Chapter 6). Further work in this direction may well open up more possibilities for new cultivars. For the curious gardener keen to test his skill, it is also an absorbing and rewarding process, always interesting and with reasonably quick results, in perhaps three years to flower, that are often productive of something really worthwhile. Since the Macrophyllae are rather genetically unstable, it is also always worth keeping a keen eye open for mutations – branch sports that produce a different and often desirable 'new' flower. 'Merveille Sanguine', 'Ayesha', 'Jean Varnier' and 'Pengwyn' are examples of popular cultivars discovered as branch sports (as illustrated earlier in Chapter 6).

These White House Farm hybrids are a good example of combining the qualities of both species.

Hydrangea scandens

I believe an influential species for creating new hybrids of different form and flower will be *H. scandens*. As its name suggests, when well grown this reaches out and into neighbouring plants with longish shoots of the season that, the following year, flower along their length from opposite pairs of nodes, which crowd at regular intervals along the shoots. These arch and bow gracefully under the weight of flower. As mentioned earlier, 'Runaway Bride', the 2018 plant of the Year at Chelsea, is a *H. scandens* hybrid showing these characters, with weather-resistant white flowers of good size from axillary buds. It will put up with full sun, remaining healthy and green, though with rather smaller flowers and leaves. Breeders no doubt will be looking to use *H. scandens* to extend the range of similar plants, but in a variety of colours. Hardiness remains to be tested, but the species itself appears fully reliable in average UK conditions.

Compatibility appears not to be a problem. As we go to press, I have just learned that one of my own *scandens* × *serrata* seedlings has evidently flowered at Tregrehan in Cornwall with all the desirable *scandens* qualities and pretty pink flowers. A new *scandens* × *macrophylla* cross called 'French Bolero' has just been introduced, also with pink flowers. My prediction of *scandens*' influence on future hybrids is already beginning to become an exciting reality.

'French Bolero', a recent exceptionally free-flowing *H. scandens* hybrid, available from nurseries already.

A promising pink White House Farm hybrid of *H. serrata* and *H. scandens*.

Hydrangea chinensis et alia

When China opened up again in 1980 to botanists and horticulturists from the West, the possibilities for the introduction of gardenworthy new species and forms were almost infinite. Apart from the collections of Ernest Wilson, there was nothing to write home about concerning hydrangeas during the golden age of plant exploration over 100 years ago. George Forrest probably trod on hydrangeas to collect more rhododendrons and primulas to satisfy demand back home, ignoring hydrangea variants of horticultural potential. Over the last 40 years, there have been many new, or perhaps more accurately, reintroductions of plants either lost to cultivation or existing only in herbaria; former species sunk into obscurity by McClintock, such as *H. davidii*, *H. stylosa* and *H. chinensis*, in a variety of forms and colours. Add in some excellent forms of *angustipetala*, *obovatifolia*, *angustifolia* and *macrosepala* introduced from Taiwan by the Wynn-Joneses, plus several from North Vietnam, and the potential for introducing valuable new blood is evident, including fragrance and, who knows, perhaps a genuine yellow hydrangea. Now that would be a real breakthrough.

Testing compatability

It has been broadly accepted that species from the different subsections of hydrangea are incompatible and, therefore, impossible to cross-fertilise to create hybrids. For example, in 1938 Mouillere crossed *H. paniculata* with a coloured form of *H. macrophylla* and, although this produced hybrid seedlings, they were not sufficiently robust to grow-on successfully.

Having said that, Michael Haworth-Booth might have been right when he claimed that this may not be an insuperable obstacle, as earlier, in 1910, a M. Foucard did succeed in breeding a hybrid between *H. macrophylla* and *H. paniculata* with pink flowers. This was exhibited in 1912, and actually received a Certficate of Merit. Unfortunately, the plant was lost during the Great War. But a race of ultra-hardy *H. paniculatas* with flowers in the full *H. macrophylla* colour range might just be possible.

In addition, Henri Cayeux, another French breeder, successfully crossed *H. macrophylla* with *H. petiolaris*, which he exhibited in 1922 under the name of *H. hortentiolaris*. The idea of a strong climbing hydrangea bearing numerous *H. macrophylla* globose flowers of rich colour simply boggles the mind. Alas, the Cayeux plants were also lost during the last war, but it demonstrates that there remains a possibility of developing potentially spectacular new plants. The possible commercial reward of a multi-million-dollar plant should make it an attractive prospect, well worth the effort for professional hybridists.

Hardy colourful climbers

Good forms of hardy climbing hydrangeas have been introduced recently by the Wynn-Joneses, among them some truly outstanding hardy forms of *H. anomola* and *H. petiolaris*. Their delightful pink 'Crug Coral' (*see* Chapter 8) would make an obvious cross with other petiolaris forms with larger flowers, or perhaps with

A selection of bright yellow *H. chinensis* forma *macrosepala* named 'Golden Crane' by Dan Hinkley. Could a yellow hydrangea be on the cards?

Schizophragma hydrangeoides roseum. © Hortival Diffusion

Crug Farm Plants' red *H. trianae.*

Schizophragma hydrangioides roseum, especially as it has now been sunk into *Hydrangea*, signalling a close relationship and therefore significant prospects for a successful cross.

The Wynn-Joneses also found plants in Columbia of *H. trianae* of the Cornidia section (*see* Chapter 9), the tree- and cliff-climbing evergreen liana hydrangeas from Latin America. *Hydrangea trianae* has dramatically rich-red ray flowers of extraordinary quality. Their plants were lost, alas, but at least their existence is now known, and it may well be possible to reintroduce them. Again, the imagination leaps at a possible cross with a hardy *H. petiolaris*, or even with the more closely-related *H. seemannii*, possibly yielding a vigorous, reliably hardy, evergreen climber, self-clinging, with intense-red, large ray flowers to light up a warm house wall. This would be a master-stroke. A cross between the Cornidia *H. peruviana* and *H. integrifolia* has already been made (*see* Chapter 9), with pink fertile flowers but no conspicuous ray flowers. This successful cross, though not greatly ornamental, suggests that a parallel cross aiming to produce a brilliant-red *H. trianae* hybrid of acceptable hardiness is not beyond the bounds of possibility.

The Wynne-Joneses at Crug Farm Plants have also introduced *H. oerstedii*, also from Columbia, with pink ray flowers. This has not been tried for hardiness and is unlikely to survive moderate winter cold but, again, crossed with another hardy climbing hydrangea species this could potentially produce a pink-flowered evergreen climber of relative hardiness. Little is known about potential compatability, but it could be tried empirically and if it worked it would certainly be a commercial winner, a veritable *chef d'oeuvre*.

Related genera now designated as hydrangeas

In Chapter 2, I described the placement into the genus *Hydrangea* of eight genera formerly in the family Hydrangeaceae, following phylogenetic analysis in the laboratory that indicated close evolutionary relationships. Given the obvious morphological differences, I questioned the value of such a change to the average gardener, who down many generations has a clear picture in his own mind and imagination of the defining characteristics of any 'hydrangea'. However, the botanists claim that one of the benefits of this exercise to gardeners is to highlight predictability in horticultural areas such as plant breeding, with a close relationship indicating probable compatibility in hybridising. It may be possible to successfully cross these plants, formerly of a different genus, with plants from the hydrangea subsection with the closest relationship to them. The distinctive characters of flower and foliage could easily yield a successful outcome, with a breakthrough in colour and, above all, flower form. Would, for example, *Schizophragma* cross with *H. macrophylla* 'Merveille Sanguine'?

One such cross was introduced by Robert Fortune in 1844, between *Dichroa* and *Hydrangea*, now known as '× *Didrangea*'. In recent years, similar hybrid selections have been made by Glyn Church, a hydrangea expert in New Zealand: one with myriad small blue flowers in quite a large corymb on a lush evergreen shrub has been named × *Didrangea* 'Cambridge Blue'. This appears to be hardy, having come through several winters, one of which killed a neighbouring *Dichroa* species.

Hydrangea macrophylla 'Merveille Sanguine' (left) and *Schizophragma* (right): a potential cross?

× *Didrangea* 'Cambridge Blue'.

Dichroa species, newly introduced from Guizhou, China.

There are other *Dichroa* species, mostly blue, which could be crossed to ma<e still new hybrids.

Other interspecific crosses have been developed and trialled in the USA and look to have very desirable garden plant characteristics.

Much of this is highly speculative stuff, but nonetheless well worth trying, and there is nothing to lose: so I hope some skilled, seasoned, competent professional will take up the challenge and attempt some of the more unlikely crosses, if they have not already done so. Enjoy the dream.

POSTSCRIPT

I asked a group of thoughtful individual hydrangea cognoscenti, a mixture of plantspeople, horticulturalists, nurserymen and judges of plant awards with no commercial axe to grind, an impossibly difficult question. Given that there are now probably something approaching 2,000 potential choices, which ten hydrangeas would they choose to grow if they were marooned on the proverbial Desert Island? Good soil conditions were a given. Of course, in practice, their choice would largely be determined by those plants with which they were most familiar, almost certainly growing in their own gardens. So this is a limited field of choice, certainly not a scientific experiment, just a fun poll.

It had to be their personal choice, no conferring was allowed and no justification for their selection was required. Their choice would be confidential, simply forming part of an interesting statistical analysis to establish which, if indeed any, cultivars or species would be preferred by a majority.

I had twelve responses, each with their ten choices.

- The two top selections, with seven votes each, were *H. macrophylla* 'Mme Emile Mouillere' and *H. macrophylla* 'Blaumeise', the old and the (relatively) new; a white 115-year-old mophead and a 50-year-old lacecap hortensia.
- With five votes each, a paniculata and a serrata came into the picture, with *H. paniculata* 'Limelight' and *H. serrata* 'Tiara' joining a third *H. macrophylla*, namely 'Dancing Snow' (syn. 'Wedding Gown').
- With four votes, enter two asperas, 'Hot Chocolate' and 'Koki'; with a fourth macrophylla, the black stemmed 'Zorro'.
- Three more asperas came in with three votes each – 'Anthony Bullivant', 'Rosemary Foster' and, rather surprisingly, *H. sargentiana*; also two paniculatas, 'Phantom' and 'Vanille Fraise'; and a macrophylla hybrid, 'Tokyo Delight'.
- Finally, with two votes each, no less than five Teller hybrids; plus votes for *H. arborescens* 'Annabelle', for two more macrophyllas, 'Générale Vicomtesse de Vibraye' and 'Europa', and a solitary quercifolia, 'Snow Queen'.

All in all a very varied mix of responses, perhaps reflecting the recent advance in popularity of species other than macrophylla, as well as the warmer winters we seem to be enjoying in the UK. The astonishing choice is that at top of the list is 'Mme Emile Mouillere', still up and running after no less than 115 years, in spite of the persistent competition from 'new' varieties. What a tribute!

BIBLIOGRAPHY

Bean, W. J. *Trees and Shrubs hardy in the British Isles* (8th edn, John Murray, London 1988)

Church, Glyn. *Hydrangeas* (Cassell, London, 1999)

Cullen, James, Sabina Knees and Janet Cubey (eds). *The European Garden Flora* (six volumes, last edition published by Cambridge University Press, 2011)

Dirr, Michael. *Hydrangeas for American Gardens* (Timber Press, OR, USA, 2004)

Dirr, Michael. *The Hydrangea Book: The Authoritative Guide* (Stipes Publishing, IL, USA, 2021)

Haworth-Booth, Michael. *The Hydrangeas* (fourth edition, Constable, London, UK, 1984)

Haworth-Booth, Michael. *Effective Flowering Shrubs* (Collins, London 1970)

Hillier, John (originator) Dawn Edwards, Roslayn Marshal *et al* (eds) *The Hillier Manual of Trees and Shrubs* (online resource at https://www.hillier.co.uk/online-shop/gifts/hillier-books/the-hillier-manual-of-trees-shrubs-2019/ Royal Horticultural Society, 2019)

Hinkley, Dan. *The Explorers Garden: Shrubs and Vines* (Timber Press, USA, 2009)

Lloyd, Christopher. *The Well-Tempered Garden* (Collins, London, 1973)

Mallet, Robert *et al*. *Hydrangea: International Index of Cultivar Names* (Edition No. 7, Association Shamrock, Varengeville-sur-Mer, France, 2018)

Mallet, Corinne (ed.). *Journal de l'Association des amis de la Collection 'Shamrock'* (available online to members in French and English: https://www.parcsetjardins.fr/jardins/417-shamrock-collection-nationale-d-hydrangea-ccvs)

Mallet, Corinne. *Hydrangea – Portraits: Portraits d'hydrangéas* (Les Editions Eugen Ulmer, Paris, 2008)

McClintock, Elisabeth. *A Monograph of the Genus Hydrangea* (Proceedings of the California Academy of Sciences, 1957)

Meier, Fritz. *Tellerhortensien-Zuechtungen* (Flugschrift Nr 120, Eidgenossische Forschungsanstalt fur Obst-,Wein-, und Gartenbau CH 8820, Wädenswil, Switzerland, 1990)

Ormezzano, Eva Boasso. *Il Libro delle Ortensie e delle Idrangee* (Libreria della natura, Milan, 2018)

Philip, Chris *et al* (originators), now Cubey, Janet (ed.) *et al*. *RHS Plant Finder* (Royal Horticultural Society, https://www.rhs.org.uk/about-the-rhs/publications/plant-finder, updated online resource with regular print editions)

Samain, Marie Stephanie. 'Hydrangea grows bigger' in *The Plant Review* (Royal Horticultural Society, December 2021)

Sargent, Charles Sprague (ed.) *Plantae wilsonianae* (Dioscorides Press, USA, 1988: first published by Cambridge University Press, 1917)

Van Gelderen, C. J. and Van Gelderen, D. M. *Encyclopaedia of Hydrangeas* (Timber Press, OR, USA, 2004)

Yamamoto, Takeomi. *The Japanese Hydrangeas Colour Guide Book* (adaptation française by Corinne Mallet, Association Shamrock, Varengeville-sur-Mer, France, 2002)

Author notes taken at 'Hydrangea 2007', the international conference held at the University of Ghent Botanic Garden

Various author articles in RHS publications, *The Garden* and *The Plant Review*

INDEX

First published in 2023 by
The Crowood Press Ltd
Ramsbury, Marlborough
Wiltshire SN8 2HR

enquiries@crowood.com

www.crowood.com

British Library Cataloguing-in-Publication Data
A catalogue record for this book is available from the British Library.

ISBN 978 0 7198 4283 2

Cover: Kisawa no hikare – a unique Japanese cultivar of *H. serrata* in both flower and foliage.
Page 2: A White House Farm deep-blue *H. serrata* seedling.
Page 4: A White House Farm *H. aspera* hybrid – 'Hot Chocolate × Titania' – with its large pink flowers enhanced by dark foliage and red/pink stems.

Typeset by Envisage IT
Cover design by Sergey Tsvetkov
Printed and bound in India by Parksons Graphics Pvt. Ltd.

Acknowledgements
My experience of hydrangeas comes from growing them and from lifelong exchanges with valued colleagues all over the world from whom I have learned a great deal over many years. However, in compiling this book my principal collaborator, supporter, critic, photographer and indispensable techie is my daughter Clare and I am deeply in her debt.

I should also, in particular, like to acknowledge the contribution of John Massey and Philip Baulk of Ashwood Nurseries for sharing freely their experience and images; Bleddyn and Sue Wynne-Jones of Crug Farm Plants who generously shared their knowledge and their time for me to study the extraordinary wealth of their many hydrangea introductions from the wild in South-East Asia in their garden and nursery; to Robert and Corinne Mallet, creators of the National Shamrock Collection in Varengeville-sur-Mer, France, whose generosity in spreading both knowledge and plants has been of enormous benefit; Roger Butler of Signature Hydrangeas at Golden Hill Nurseries in giving me time and facility to photograph some of his most recent acquisitions, and finally my warmest thanks to Midori Shiraishi in Japan, whose manifest love of the plants is equalled only by the depth of her knowledge of their history in Japanese culture, which she has shared so generously.

Others who also kindly filled in gaps in my photo library include: The Royal Horticultural Society; Chris Sanders; Trish Fry, and Bob Howard in Nova Scotia; Dirk and Cor van Gelderen at the Esveld Nursery in Holland; Everard Daniel; Kevin Hobbs; Caradoc Doy; and Jukka Kallijarvi in Finland. My warm thanks to them.

Image credits
Ashwood Nurseries, p.15 (bottom), p.109 (top left); Bob Howard, p.45, p.118 (top right); Bleddyn Wynn Jones, p.159, p.199 (bottom); Caradoc Doy, p.14 (top); Chris Sanders, p.44 (top and bottom), p.48 (all), p.52 (bottom left), p.101 (top right); Everard Daniel, p.103 (right-hand column, middle), p.109 (bottom left), p.110 (top right), p.112 (left-hand column, middle; right-hand column, top); Hortival Diffusion, p.192, p.199 (top); Jukka Kallijarvi, p.40 (top); Kevin Hobbs, p.43 (bottom right); Midori Shiraishi, p.19, p.89 (left), p.116 (bottom), p.117 (left-hand column, top and bottom), p.166 (top left and top right); Robert Mallet, p.96 (all), p.98 (top left); Royal Horticultural Society, p.50 (all); Trish Fry, p.52 (top left); Van Gelderen, p.98 (second down, left-hand column), p.99 (second down, left-hand column; bottom, left-hand column; second down, right-hand column; third down, right-hand column; second from bottom, right-hand column; bottom, right-hand column), p.103 (middle pair, bottom), p.109 (left-hand column, bottom; right-hand column, middle and bottom).